INVESTIGATIONS IN NUMBER, DATA, AND SPACE

Implementing the *Investigations in Number, Data, and Space®* Curriculum

Grades 3–5

Susan Jo Russell
Cornelia Tierney
Jan Mokros
Anne Goodrow
Megan Murray

Developed at TERC, Cambridge, Massachusetts

Dale Seymour Publications®

The *Investigations* curriculum was developed at TERC (formerly Technical Education Research Centers) in collaboration with Kent State University and the State University of New York at Buffalo. The work was supported in part by National Science Foundation Grant No. MDR-9050210. TERC is a nonprofit company working to improve mathematics and science education. TERC is located at 2067 Massachusetts Avenue, Cambridge, MA 02140.

This project was supported, in part,
by the
National Science Foundation
Opinions expressed are those of the authors
and not necessarily those of the Foundation

This book is published by Dale Seymour Publications®, an imprint of the Alternative Publishing Group of Addison-Wesley Publishing Company.

Managing Editor: Catherine Anderson
Series Editor: Beverly Cory
Consulting Editor: Priscilla Cox Samii
Manuscript Editor: Mali Apple
ESL Consultant: Nancy Sokol Green
Production/Manufacturing Coordinator: Barbara Atmore
Design Manager: Jeff Kelly
Cover: Bay Graphics

Copyright © 1997 by Dale Seymour Publications®. All rights reserved. No part of this publication may be reproduced in any form or by any means without prior written permission of the publisher. Printed in the United States of America.

 Printed on Recycled Paper

Order number DS21663
ISBN 1-57232-380-9

3 4 5 6 7 8 9 10-ML-00 99 98 97 96

CONTENTS

Grade 3 Components of *Investigations in Number, Data, and Space* — 1

Grade 4 Components of *Investigations in Number, Data, and Space* — 2

Grade 5 Components of *Investigations in Number, Data, and Space* — 3

The Changing Face of Mathematics Instruction — 4

Organization of the *Investigations* Curriculum — 5
  Units
  Investigations
  Sessions
  Extensions and Excursions

The Mathematics Content of the *Investigations* Curriculum — 7
  Number
  Data
  Geometry
  Change

Special Features of the *Investigations* Curriculum — 9
  Materials and Technology
  Cooperative Learning and Grouping
  Assessment
  Family Connections
  Review and Practice

The Teachers' Role — 13
  Meeting Students' Needs
  Teacher Support

Teaching Mathematics to *All* Students — 15
  Investigating Real-Life Contexts
  Investigating Mathematical Contexts

Getting Started with the *Investigations* Curriculum — 17
  Making Choices for Your School or System
  Replacement Unit Strategy
  Grade Levels in the *Investigations* Curriculum

The *Investigations* Curriculum and the NCTM Standards — 19
  Standard 1: Mathematics as Problem Solving
  Standard 2: Mathematics as Communication
  Standard 3: Mathematics as Reasoning
  Standard 4: Mathematical Connections
  Grade 3 Correlation to NCTM Standards 5–13
  Grade 4 Correlation to NCTM Standards 5–13
  Grade 5 Correlation to NCTM Standards 5–13

GRADE 3 COMPONENTS OF *INVESTIGATIONS IN NUMBER, DATA, AND SPACE*

Grade 3 Units	Pacing	
Mathematical Thinking at Grade 3 (Introduction)	3–4 weeks	
Things That Come in Groups (Multiplication and Division)	4–5 weeks	
Flips, Turns, and Area (2-D Geometry)	2–3 weeks	(disk included; 💻 optional)
From Paces to Feet (Measuring and Data)	3–4 weeks	
Landmarks in the Hundreds (The Number System)	3–4 weeks	
Up and Down the Number Line (Changes)	3–4 weeks	
Combining and Comparing (Addition and Subtraction)	3–4 weeks	
Turtle Paths (2-D Geometry)	3–4 weeks	(disk included; 💻 required)
Fair Shares (Fractions)	3–4 weeks	
Exploring Solids and Boxes (3-D Geometry)	3–4 weeks	

Support Materials

Implementing the Investigations in Number, Data, and Space® Curriculum
Family Letters (in English, Spanish, Vietnamese, Cantonese, Hmong, Cambodian)
Ten-Minute Math
The Investigations Curriculum: Bringing Together Students, Teachers, and Mathematics

Teacher Resource Package

Overhead transparencies
Blackline masters
Array cards
Elementary rulers
Fraction dice
Numeral cards
Building straws
Pad of 100 charts
Pads of graph paper (various sizes)

Materials Kit

Centimeter cubes
50-centimeter rulers
Geometric models (wooden)
Inch cubes
Number cubes
Pan balances
Pattern blocks (wooden)
Play money (coins)
Interlocking cubes
Color tiles
Adding machine tape

GRADE 4 COMPONENTS OF *INVESTIGATIONS IN NUMBER, DATA, AND SPACE*

Grade 4 Units	Pacing	
Mathematical Thinking at Grade 4 (Introduction)	3–4 weeks	
Arrays and Shares (Multiplication and Division)	3–4 weeks	
Seeing Solids and Silhouettes (3-D Geometry)	3–4 weeks	
Landmarks in the Thousands (The Number System)	3–4 weeks	
Different Shapes, Equal Pieces (Fractions and Area)	2–3 weeks	
The Shape of the Data (Statistics)	3–4 weeks	
Money, Miles, and Large Numbers (Addition and Subtraction)	3–4 weeks	
Changes Over Time (Graphs)	3–4 weeks	
Packages and Groups (Multiplication and Division)	3–4 weeks	
Sunken Ships and Grid Patterns (2-D Geometry)	2–3 weeks	(disk included; ▢ required)
Three out of Four Like Spaghetti (Data and Fractions)	2–3 weeks	

Support Materials
Implementing the Investigations in Number, Data, and Space® Curriculum
Family Letters (in English, Spanish, Vietnamese, Cantonese, Hmong, Cambodian)
Ten-Minute Math
Beyond Arithmetic

Teacher Resource Package
Overhead transparencies
Blackline masters
Array cards
Elementary rulers
Fraction dice
Numeral cards
Pad of geoboard dot paper
Pad of 100 charts
Pads of graph paper (various sizes)

Materials Kit
Centimeter cubes
Geoboard (wooden)
Geometric models (wooden)
Lima beans (dried)
Measuring tapes
Pattern blocks (wooden)
Play money (bills and coins)
Interlocking cubes
Color tiles
Adding machine tape
Centimeter rulers

GRADE 5 COMPONENTS OF *INVESTIGATIONS IN NUMBER, DATA, AND SPACE*

Grade 5 Units	Pacing	
Mathematical Thinking at Grade 5 (Introduction, Landmarks in the Number System)	4–5 weeks	
Picturing Polygons (2-D Geometry)	3–4 weeks	(disk included; required)
Name That Portion (Fractions, Percents, Decimals)	5–6 weeks	
Between Never and Always (Probability)	2–3 weeks	
Building on Numbers You Know (Computation and Estimation Strategies)	6–7 weeks	
Measurement Benchmarks (Estimating and Measuring)	2–3 weeks	
Patterns of Change (Tables and Graphs)	3–4 weeks	(disk included; optional)
Containers and Cubes (3-D Geometry: Volume)	3–4 weeks	
Data: Kids, Cats, and Ads (Statistics)	3–4 weeks	

Support Materials
Implementing the Investigations in Number, Data, and Space® Curriculum
Family Letters (in English, Spanish, Vietnamese, Cantonese, Hmong, Cambodian)
Ten-Minute Math
The Investigations Curriculum: Bringing Together Students, Teachers, and Mathematics

Teacher Resource Package
Overhead transparencies
Blackline masters
Numeral cards
Measuring prism
3-D patterns
Cat poster
Many Squares poster
Pad of One Million Dots paper
Pad of 100 charts
Pads of graph paper (various sizes)

Materials Kit
Interlocking cubes
Centimeter cubes
Color tiles and overhead color tiles
Elementary bar mass set
Blank dice with stick-on dot labels
Measuring pitchers: liter, quarts, pint, and cup
Meter sticks
Power Polygons
Spectrum school balance
Transparent blank spinners

THE CHANGING FACE OF MATHEMATICS INSTRUCTION

Mathematics classrooms across the United States are changing, and changing drastically. The focus is shifting toward an emphasis on mathematical reasoning and problem solving in a true sense. The *Curriculum and Evaluation Standards for School Mathematics* (National Council of Teachers of Mathematics, 1989) and other current reform documents emphasize that to solve problems, students must learn to describe, to compare, and to discuss their approaches. Alternative strategies are valued, multiple strategies are encouraged, and communication about mathematics is central.

Compare the notable features of the old style of elementary mathematics classroom and the class environment many educators are now striving to create.

In the old-style mathematics class, students	*In the new mathematics class, students*
• worked alone	• work in a variety of groupings—as a whole class, individually, in pairs, and in small groups
• focused on getting the right answer	• consider their own reasoning and the reasoning of other students
• recorded by only writing down numbers	• communicate about mathematics orally, in writing, and by using pictures, diagrams, and models
• completed many problems as quickly as possible	• thoughtfully work on a small number of problems during a class session, sometimes working on a single problem for one or several sessions
• used a single, prescribed procedure for each type of problem	• use more than one strategy to double-check
• used only pencil and paper, chalk and chalkboard as tools	• use cubes, blocks, measuring tools, calculators, and a variety of other materials

Many elementary school teachers are eager to change their classroom practices in order to engage their students more deeply in mathematics. What they need—whether they are new to the profession or seasoned veterans—are curriculum materials that will help them learn more about the shifting emphases toward mathematical reasoning and problem solving, even as they are teaching.

The *Investigations in Number, Data, and Space*® curriculum meets that need. It is a learning tool for teachers as well as for students. As teachers use the curriculum in their classrooms, they have at their fingertips helpful information about what they are teaching and the ways students are learning.

This implementation guide will introduce you to the *Investigations* curriculum, covering

- how the *Investigations* curriculum is structured
- how to get started in the program
- how to take advantage of the choice and flexibility the program offers in tailoring the curriculum to your own school or system
- how the program supports new roles for teachers and students
- how the program can help all students learn significant mathematics content as they begin to think and reason mathematically

ORGANIZATION OF THE *INVESTIGATIONS* CURRICULUM

Investigations in Number, Data, and Space is a K–5 mathematics curriculum that looks and feels quite different from the traditional elementary mathematics program. While it provides all the information teachers need to implement a complete mathematics curriculum, there are no student textbooks.

Units

The curriculum at each grade level is organized into *units*. Each unit offers from two to six weeks of mathematical work on topics in number, data analysis, and geometry. Because of the many interconnections among mathematical ideas, units may revolve around two or three related areas—for example, fractions and area, multiplication and division, measuring and data.

In each unit, students explore the central topics in depth through a series of *investigations*, gradually encountering and using many important mathematical ideas. Rather than working through a textbook or workbook, doing page-by-page exercises, students are actively engaged in working with materials and with their peers to solve larger mathematical problems.

The main teaching tool is a single resource book, called the *teacher book,* for each unit in a grade level. Each teacher book provides

- lesson plans
- materials lists
- reproducible student sheets for activities and games
- a letter to introduce the curriculum to each student's family
- homework suggestions
- opportunities for skill practice
- assessment activities
- notes to the teacher about the mathematics students are encountering
- classroom dialogues based on what happened in field-test classrooms

On pages 1–3 is a list of the *Investigations* units for grades 3–5, along with other components of the program. (The first complete levels available for classroom use are grades 3–5. For information about the publication dates of units for grades K–2, contact the publisher.)

Investigations

Each unit in the curriculum is built around several investigations, which offer a variety of problem contexts for students to explore. Investigations vary in length; some will take two or three days to complete, others a week or even two weeks. The investigations within each unit are designed for use as a cohesive block rather than as isolated projects; although their contexts might be quite different, they work together in a carefully planned way.

For example, in the unit *Combining and Comparing* (grade 3), one investigation has students collecting data about the ages of oldest relatives, the sizes of pets, and the number of people in a family, and comparing their personal data to world record data. In a later investigation in that unit, students work with the calendar, finding "how much longer?" to significant dates in their lives, and comparing the number of days children go to school in different countries around the world. The link between these two investigations, and among other investigations in the unit, is the creation of problem situations for students who are developing understanding of and strategies for addition and subtraction.

You will see several different types of investigations in this curriculum.

- Some investigations are structured around a set of related problems, such making geometric shapes on coordinate grids both on paper and on the computer in *Picturing Polygons* (grade 5).

- In other investigations, students use mathematical relationships as they build or construct something—such as an array of fraction number lines in *Name That Portion* (grade 5).

- Some investigations are based on games designed to involve students in thinking about particular mathematics, such as the games played with arrays in *Things That Come in Groups* (grade 3).

- Many investigations involve students in collecting and representing data, as in *The Shape of the Data* (grade 4), in which students measure and compare the heights of first and fourth graders in their school.

- Some investigations, usually at the end of a unit, are projects in which students use and extend the unit's mathematical ideas. For example, in the final investigation of *Exploring Solids and Boxes* (grade 3), students create a model city from box-patterns they draw on graph paper.

What all these different types of investigations have in common is that *students work in depth on a small number of problems,* actively using mathematical tools and consulting with peers as they find their own ways to solve the problems. Significant time is allowed for students to think about the problems and to model, draw, write, and talk about their work.

Through this type of curriculum, students better understand, enjoy, and appreciate mathematics. When students are involved in designing a toy from interlocking cubes and then writing their own directions for building it, or playing a game that involves relationships among numbers up to 10,000, or bringing their knowledge of the environment to a study of how quickly plants grow, they recognize their work as *serious problem solving with a purpose.* They develop a sense of appreciation for the power and beauty of mathematics as they learn to value their own thinking and strategies.

Sessions

Each investigation is divided into several class *sessions.* A session is a *one-hour* mathematics class.

The structure of the sessions is different from a typical mathematics lesson. Rather than have the teacher begin the period by explaining the mathematics, demonstrating techniques, and talking students through the first problem, the bulk of the whole-group time is at the end of the session, when students are comparing their methods and results, analyzing their work, and sharing conclusions.

Sessions are often grouped together to reflect the flow of the activities as they actually happen in classrooms. That is, an investigation might be presented in separate chunks—say, Session 1, Sessions 2–3, and Sessions 4–6—that reflect the continuity of the activities. Sessions are connected; they flow into each other as students' work builds and grows. In the above sequence, when Session 3 begins, students may be continuing their work on a problem they began in Session 2. There is no need for teachers to present the problem or assignment; students simply find their group, organize their materials, and continue working. With this structure, teachers spend less time presenting and more time circulating throughout the classroom, interacting, observing, listening, and questioning.

The teacher books offer ways to structure the daily work. As teachers become more familiar with the units, they will want to make their own decisions about pacing and structure to organize the work to best meet the needs of a particular class.

Extensions and Excursions

Each unit contains more than enough material for the two to six weeks it is intended to take. In addition to the basic activities for each one-hour session, some sessions include suggestions for Extensions.

Extensions provide problem situations for students to further explore the mathematical ideas in those sessions or present ways in which teachers can connect the mathematics work to other areas of class study or student interest. Extensions offer the chance for students who are particularly engaged and interested in an area of mathematics to pursue that interest.

In addition, some units have sessions designated as Excursions. These "side trips" provide additional, related activities teachers may choose to pursue with the whole class.

Extensions and Excursions provide teachers with the flexibility to tailor the curriculum to the needs, interests, and experiences of the students in their classroom, to make connections with other areas of the curriculum, and to pursue their own intellectual curiosity.

THE MATHEMATICS CONTENT OF THE *INVESTIGATIONS* CURRICULUM

The title of the program, *Investigations in Number, Data, and Space,* reflects the view that mathematics in the elementary school is more than arithmetic. In the elementary grades, students need to develop a foundation in several key content areas of mathematics: work with number, work with data, and work with geometry (space). The *Investigations* curriculum also includes work based on recent research on young children's work with the mathematics of change.

Number

One of the concerns many people have about new mathematics programs is that students will not learn to add, subtract, multiply, and divide. A central objective of the *Investigations* program is to support students' learning about number, number relationships, the base-ten number system, and number operations.

Third through fifth grades are a critical period for developing students' number and operation sense. The *Investigations* program spends a large percentage of time on this area, with a focus on the development of students' own strategies for solving problems. Fluency and accuracy *are* critical. Students *do* need to learn their addition and multiplication "facts," and many *Investigations* activities will help them do so. However, speed should not be confused with fluency, nor memorization with building a flexible knowledge of number relationships.

In the *Investigations* curriculum, students develop sound strategies they can rely on to solve computation problems. They learn to look at the whole problem and make reasonable estimates of the result. They use materials and models to visualize the relationships of quantities in addition, subtraction, multiplication, and division situations. They have experience working with calculators and other mathematical tools. They keep track of their work by recording intermediate steps of a problem. And they learn to have more than one strategy to solve any problem so that they can double-check their accuracy.

Work with number is not limited to work with operations. Students should understand number as a way to describe relationships in the real world, but they should also encounter purely mathematical questions (What patterns can you find in a 10 by 10 array of the numbers 1 to 100? What are the factors of 1000? 1100? What happens when you add two odd numbers together? How might 13×10 help you to solve $159 \div 13$?). Number relationships are in themselves fascinating objects of study. In the *Investigations* units, students have the opportunity to experience, appreciate, and be fascinated by the patterns of number and to make and test conjectures based on what they see.

Data

Data collection and analysis is a critical skill in an information-rich society. From the earliest grades, students can collect, display, describe, and interpret real data so they learn to become critical users of data and graphs. In the *Investigations* curriculum, students pose their own questions, collect data, critique and refine their own data-collection methods, compare different ways of displaying their data, and begin to use statistical terms and measures (range, median, outlier).

Research on students' understanding of statistical ideas in the elementary grades indicates that, just as in work with number, focus on memorization of definitions and algorithms (such as the algorithm for calculating the mean) undermines students' learning to make sense of a set of data. Just as students pull numbers out of word problems and manipulate them blindly, they pull numbers out of data sets and carry out calculations that no longer have meaning to them in terms of the data. The *Investigations* curriculum includes many opportunities for students to describe, analyze, and interpret a variety of data sets so they begin to understand how data analysis can provide important information about a variety of populations.

Geometry

Geometry, as the study of spatial objects, relationships, and transformations, is an essential component of mathematics. Through the study of measurement, geometry serves as a major source of practical applications of numerical concepts.

Investigations in geometry and measurement provide opportunities for students to mathematically analyze their spatial environment, to describe characteristics and relationships of geometric objects, and to use number concepts in a geometric context. In their studies of geometry, students encounter polygons, polyhedra, symmetry, and geometric motions. They measure length and distance; perimeter, area, and volume; angles; weight and capacity. They use a variety of strategies to locate themselves and other objects in space, on maps, or on coordinate grids.

Much of the thinking required in higher mathematics is spatial in nature. Many believe that some people are just naturally good at spatial visualization, while others are not. On the contrary, as with most other skills, children and adults become better at spatial visualization by spending time doing activities that require it. The 3-D geometry units in the *Investigations* curriculum emphasize spatial visualization as well as the construction and description of three-dimensional objects. As students develop ways of visualizing relationships in space, they can more readily incorporate the use of spatial models to support their understanding of number relationships.

Change

Change is one of the most pervasive aspects of our lives. We are constantly experiencing the flow of time and the changes that occur over time—motion, growth, and temperature, for example. One of the driving forces of mathematics is to understand and predict change. Units emphasizing aspects of the mathematics of change (for example, *Up and Down the Number Line* in grade 3, *Changes Over Time* in grade 4, and *Patterns of Change* in grade 5) include activities in which students learn to construct and interpret graphs of changing populations, plant growth, geometric patterns, and motion. They interpret graphs as stories of changes over a time period. They use sequences of addition and subtraction to describe situations of discrete change. Because the mathematics of change is often used to describe physical situations involving growth and time, integration between science and mathematics is an important focus of these investigations.

Many of the ideas developed in these units are central to the continued study of mathematics in middle and high school. Research on students' understanding of mathematics indicates that the core ideas of "advanced" mathematics develop as enrichment and refinements of basic intuitions that children express from a very early age. Work in the *Investigations* curriculum emphasizes qualitative understanding of these ideas as students discuss the meaning of graphical and numerical patterns. The units strengthen the continuity between elementary school mathematics and "advanced" courses such as algebra and calculus by introducing important ideas about growth and change.

SPECIAL FEATURES OF THE *INVESTIGATIONS* CURRICULUM

Materials and Technology

Mathematics Materials and Manipulatives

Concrete materials are central to the *Investigations* curriculum. Some of the materials used are mathematical tools, such as interconnecting cubes, 100 charts, geometric shapes, play money, and rulers. Others are specific to particular problem situations, such as a small toy figure to be a "viewer" in a geometric landscape or dried beans for comparing handfuls. Common classroom supplies such as chart paper, scissors, crayons or markers, glue or gluesticks, and various kinds of paper are often needed.

Basic mathematics manipulatives, such as interconnecting cubes, measuring tools, and 100 charts, should always be readily available to students. Students must know how to take materials out, how to use them, and how to return them to storage. The introductory units—*Mathematical Thinking at Grade 3, Mathematical Thinking at Grade 4,* and *Mathematical Thinking at Grade 5*—provide opportunities for students to become familiar with using and caring for basic mathematical materials.

A key part of the teachers' job is to ensure that it becomes natural for students to use appropriate materials as they solve problems. At first, teachers may need to put buckets or bins of materials they know will be useful on tables, within reach of each group of students. Later, students might be encouraged with specific suggestions: "Will you be using the cubes or the coins? . . . OK, get a bucket of cubes from the math shelves." As the year progresses, students will become accustomed to choosing appropriate materials as a regular part of their mathematical work.

Each unit contains a complete list of materials needed for that unit. The curriculum was designed to balance the need for students' access to materials to adequately explore mathematics and the restrictions on resources many schools face. Where possible, suggestions are given for alternative materials, ways to make or duplicate games and charts, and inexpensive materials that teachers or students may be able to bring from home.

All materials that must be reproduced, such as student recording sheets, homework assignments, problem cards, and games, are included as blackline masters in the back of each unit. Classroom pads of these reproducible materials are available, and a classroom kit of basic manipulative materials is available for each grade level. However, teachers may want to put together classroom materials by purchasing only what is needed to supplement what is already available at their school.

An overhead projector is recommended in all units. It provides a visual point of focus for the whole class as teachers pose problems visually or as students share their strategies, approaches, and questions. Blackline masters for transparencies are provided at the back of each unit. Transparencies for each grade level are also available separately.

Calculators

Calculators are used in many units. Ideally, they will always be available in the classroom and, like other mathematics materials, students will use them as needed. Students need to become well acquainted with the calculator so they can use it fluently and flexibly as one method for solving numerical problems. Students need a repertoire of strategies for work with number operations—including mental arithmetic and estimation, use of concrete materials, recording with pencil and paper, and calculators. Through the curriculum, students will learn to use each of these approaches separately and in combination (for example, it is often useful to jot down intermediate results with pencil and paper as you solve a problem mentally).

Many activities throughout the units require the use of the calculator. In some activities, students are asked to use other strategies and not to use the calculator. In many situations, students are required to solve a problem in more than one way; use of a calculator might be one of the ways.

Computers

In several units, students use computers to explore geometry, measurement, and coordinate grids. One unit at each grade level—*Turtle Paths* (grade 3), *Sunken Ships and Grid Patterns* (grade 4), and *Picturing Polygons* (grade 5)—requires the use of computers. The use of computers is advantageous for *Flips, Turns, and Area* (grade 3) and *Patterns of Change* (grade 5) but not required. Each of these

Special Features of the Investigations *Curriculum* ■ 9

units includes a disk of special activities created for the *Investigations* curriculum using *Geo-Logo*™, a version of the Logo language designed for the study of geometry in the elementary classroom. (The disk requires a Macintosh II or above with 4 MB of internal memory and Apple System Software 7.0 or later.) The units are written to be carried out either with a small number of computers in the classroom or in a computer lab. Each of these units integrates off-computer and on-computer activities so students can explore the mathematical content in a variety of contexts and so teachers can organize classroom activities in accordance with the amount of computer access available to them.

Cooperative Learning and Grouping

In an *Investigations* classroom, students work individually and in a variety of heterogeneously mixed groupings—pairs and small groups of three or four—and as a whole class.

Students Work Cooperatively

Students spend much of their time in mathematics class in pairs or small groups. By working with other students to solve problems, students are encouraged to articulate their ideas and to compare their ideas to those of their peers. They benefit both from explaining or demonstrating their own approaches to a mathematics problem and from seeing solutions that may be quite different from their own.

As students become used to an atmosphere in which problems are solved cooperatively, with all students' contributions being valued, they become more focused on thinking about the mathematics rather than on concealing their work from their classmates.

Students Communicate About Mathematics

Talking as they work is an important aspect of students' experience in the mathematics classroom. As a pair of students evaluates their work together, they are more likely to think their way out of dead ends and to recognize false paths on their own. Often, one student, sparked by another student's comment, makes a new connection or discovers an important next step. As students learn to work together, they tend to make even "winning" games collaborative, giving each other suggestions about moves and strategies.

Teachers Facilitate Cooperative Work

Teachers need to find ways to support cooperative work in their classrooms. For example, the physical arrangement of the classroom must provide ways for students to work and talk together easily. Many teachers arrange desks in groups of four; others use small tables or create comfortable areas where students can work together on a rug.

Who works together and for how long is a practical problem that takes thought and planning. Teachers who are skilled at cooperative grouping use a variety of strategies; for example, while they are establishing classroom rules and values about working together, teachers choose the members of pairs and small groups so students become used to working with a variety of partners. As students become more skilled with collaborative work, many teachers use a mixture of student and teacher choice for groupings.

Talking with students about collaborative work is a critical part of the cooperative-learning process. Students learn to voice their opinions about how well groups are working, how to resolve conflicts, how to help everyone contribute to the group, and how to allow for individual differences in working style. It is important for teachers to acknowledge that some students will need time to pursue their ideas individually as well as in their work in pairs or small groups.

Small-group and individual work times are punctuated by whole-class gatherings for discussions of questions students have raised or comparisons of problem strategies and solutions. The teacher books provide guidelines for these discussions, including examples of questions teachers or students might raise and hints for how to begin the discussion and questions to keep it going. Dialogue Boxes in each unit give examples of how some of these discussions have unfolded in real classrooms.

A critical role for teachers is to foster participation by all students over the course of the year in these group discussions. By observing and listening carefully during small-group work time, teachers can identify students who have important questions or ideas to share with the whole group. Several times during a unit's work, teachers can carefully choose one or two student examples to share with the class so students who don't volunteer their work are included and recognized.

Assessment

Each unit includes several kinds of assessment. From the students' point of view, these assessments are no different from other curriculum activities; they don't look or feel like traditional tests or end-of-chapter quizzes. However, each assessment provides teachers with ways to reflect on student understanding.

- **Teacher Checkpoints** provide a time for teachers to pause and reflect on their teaching plans while observing students at work. These sections offer suggestions about what to look for, what questions to ask, and how to modify instruction or pacing to meet students' needs.
- **Embedded assessment activities** help teachers examine the work of individual students, figure out what it means, and provide feedback. These activities sometimes involve writing and reflecting, at other times a brief interaction between student and teacher, and in others the creation and explanation of a product.
- **Ongoing assessment** of students throughout the year supports the assessment activities. Observing and listening to students is one of the key jobs of teachers as they continually try to refine their understanding of how their students understand mathematics. The use of portfolios or mathematics notebooks to help document student work during the course of a year is encouraged. The final activity in each unit is "Choosing Student Work to Save," which provides suggestions for selecting student work, having students write summary comments about their work in the unit, and sharing work with families. Selecting and compiling student work over the course of a year serves several purposes. Teachers are able to keep track of the development of students' mathematical thinking. By examining a selection of work, teachers, students, and family members can evaluate student progress together. In the course of selecting and writing about their work, students have an opportunity to reflect on the quality of their work and what they have learned in each unit.

Family Connections

Family Letters

Just as it is important for children to read at home, it is essential for them to do mathematics at home and to involve their families in this enterprise. Families, like students, need to understand that figuring out how to solve an actual problem—not memorizing the answer—is the essence of mathematics. This is a new perspective for many families.

In recognition of this, the *Investigations* curriculum includes informative letters that introduce families to the mathematics their children are doing in each unit. Each letter gives families an idea of what their children are doing in class, explains how this is connected to the rest of their children's lives, describes some of the homework children will be doing, and provides ideas for families about how to expand on the mathematics that is being done at school. Family letters are available in six languages: English, Spanish, Vietnamese, Cantonese, Hmong, and Cambodian.

Family Meetings

Some teachers find that the intent and organization of the *Investigations* curriculum is easier for families to assimilate if they are invited to a special meeting and formally introduced to the new program. During this meeting, families can see the *Investigations* activities and materials in action as they experiment with a few themselves, ask questions about what and how their children will be learning, and experience for themselves how engaging mathematics class can be.

Homework

In the *Investigations* curriculum, homework is a vehicle for connecting school mathematics with students' everyday lives. Homework assignments often involve gathering data at home. For example, students may be asked to make a representation of how many people are in their home over a 24-hour period, to check on the prices of food advertised in the local newspaper, or to chart the distance from their stove to their front door.

Students are frequently asked to work on problems that extend and solidify their mathematical understanding. For example, after they have worked on a similar problem in class, students may be asked to make a collection of things at home that contains close to 1000 items. At other times, they will bring home number games, teach family members how to play, and then engage in a few rounds of friendly competition. These games have the dual purpose of giving students practice in thinking about number operations and relationships and giving their families a glimpse of a more exciting and meaningful way to learn "traditional" mathematics.

Review and Practice

Each unit in the *Investigations* curriculum explores an area of mathematical content in depth. For example, *Landmarks in the Hundreds* (grade 3) explores the base-ten number system; *The Shape of the Data* (grade 4) offers investigations in data analysis; *Name That Portion* (grade 5) provides problems with fractions, decimals, and percents; *Seeing Solids and Silhouettes* (grade 4) gives work in spatial visualization; and *Things That Come in Groups* (grade 3) deals with multiplication and division. The *Investigations* curriculum has been designed to ensure that ideas encountered in one unit recur and that skills used during a particular unit are reviewed and practiced.

Many of these ideas and skills occur in several units, even when they are not the unit's main focus. For example, data analysis is not the emphasis of the addition and subtraction unit, *Combining and Comparing* (grade 3), but students collect and represent data in that unit. Multiplication and division are not the core of the geometry unit *Cubes and Containers* (grade 5), but multiplication ideas underlie students' building and analyzing rectangular solids in that unit. In the unit *Different Shapes, Equal Pieces* (grade 4), geoboards—a geometric model—are used to explore fraction relationships. In *Three out of Four Like Spaghetti* (grade 4) and *Data: Kids, Cats, and Ads* (grade 5), students use fractions to report and compare data.

Ten-Minute Math Activities

Connections among mathematical content pervade the *Investigations* curriculum. However, there are some skills students need to work with for short periods of time on a regular basis. Students should be thinking about strategies for estimation and mental arithmetic throughout the year. They should have opportunities to return to work in spatial visualization that is introduced in the 3-D geometry units. Spur-of-the-moment data collection that connects to students' interests or classroom events is necessary, even if students are not actually doing a data unit.

To support students' work on these skills—estimation, mental arithmetic, data representation and analysis, spatial visualization—throughout the school year, Ten-Minute Math activities are incorporated into each unit. These activities provide review and practice of important skills and fit into any odd 10- or 15-minute time slot (outside of mathematics class) teachers have available. One or two Ten-Minute Math activities are specifically incorporated into each unit, so teachers have a structure and schedule for including them in the year's work. As teachers and students develop a repertoire of these short but substantive activities, the activities become a natural part of ongoing class activity. Most Ten-Minute Math activities include homework options for continued practice.

Support for Spanish-Speaking Students

Support materials for classrooms with Spanish-speaking students enhance the effectiveness of the *Investigations* curriculum.

The *Spanish Vocabulary Package*, available at each grade level, contains Spanish versions of all black-line masters. It also includes the *Spanish Vocabulary Booklet*, which offers Spanish translations of the material in the "Vocabulary Support for Second-Language Learners" and the "Preview for the Linguistically Diverse Classroom" sections of the *Investigations* curriculum.

The *Spanish Teaching Companion*, also offered at each grade level, is a set of booklets that provides Spanish translations of the key passages in the teacher books that are indicated by boldface type.

THE TEACHERS' ROLE

Just as the *Investigations* curriculum supports students to be actively engaged in mathematics, it is also designed to be used by teachers who are actively listening to, observing, questioning, and facilitating the mathematical work of their students. For teachers to actively participate in modifying the curriculum to work best for their students, they must become careful observers and listeners, continually trying to understand how their students are thinking about important mathematical ideas. The teachers' role is

- to observe and listen carefully to students
- to try to understand how students are thinking
- to help students articulate their thinking, both orally and in writing
- to establish a classroom atmosphere in which high value is placed on thinking hard about a problem
- to insist that students keep track of their work and be able to explain or show their thinking
- to ask questions that push students' mathematical thinking further
- to facilitate class discussions about important mathematical ideas
- to make decisions about how to modify the curriculum appropriately for individuals or groups of students

Meeting Students' Needs

Students enter the *Investigations* program, as they do any curriculum, with vastly different backgrounds, knowledge, and experience. They differ in what they believe mathematics is and how they see themselves as mathematics learners. While the *Investigations* curriculum has been developed to support the learning of a wide range of students, a critical part of the teacher's role is the continual monitoring of student learning and modification of the curriculum to meet the needs of the variety of students in the class. In particular, teachers may need to modify activities for students who come into class with poorly developed number sense, who have not developed ways to organize and keep track of their work, or who have low self-confidence in their ability to think mathematically. Many of the activities are constructed so teachers may easily adjust them, for example, by

- changing the numbers in a problem to make the problem more accessible or more challenging for particular students
- repeating activities with which students need more experience
- rearranging pairs or small groups so students learn from a variety of their peers

Teacher Support

Active mathematics teaching requires that teachers think hard themselves about the mathematics content their students are learning and about the ways in which students best learn that content. This job is difficult, requiring that teachers continually improve their own understanding of mathematics teaching and learning.

The *Investigations* curriculum was created to be a tool for continuing professional development. Several components of each unit are designed to support teachers as they reflect about teaching and learning mathematics:

- **About the Mathematics in This Unit.** This section describes the mathematical center of the unit and the connections among the key mathematical ideas students will encounter.
- **Teacher Notes.** Throughout each unit, Teacher Notes provide practical information about the mathematics content and how students learn it. Many of the notes were written in response to actual questions from teachers or to discuss important things that happened in field-test classrooms. For example, some teachers and students hold a rigid notion of mental computation, believing that if they are computing mentally, nothing must be written down. On the contrary, field tests show that when doing mental computation, people often jot down notes to themselves to keep track of their procedures or intermediate results. A Teacher Note, Keeping Track of Addition and Subtraction, provides examples of this mix of mental computation and note-taking and suggests ways teachers can model for students how to keep track of their thinking. Teacher Notes also offer teachers help with thinking

about mathematical ideas that may be unfamiliar to them and information about how these ideas support more advanced mathematics.

- **Dialogue Boxes.** Sample dialogues throughout the unit demonstrate how students typically express their mathematical ideas, what issues and confusions arise in their thinking, and how some teachers have guided particular class discussions. For example, in the unit *Turtle Paths* (grade 3), several dialogues are devoted to students' confusion as they begin to work with the idea of *angle* and to the ways in which teachers can support students' thinking as they struggle with these new ideas. The dialogues are based on the extensive classroom testing of the *Investigations* curriculum. They offer good clues to how students may develop and express their approaches and strategies, and ways of guiding class discussions.

These teacher-support components allow teachers on their own or, even better, with a group of peers in their school or system, to refamiliarize themselves with mathematical ideas, to learn new mathematics, and to reflect deeply on how to support students as they learn. Ultimately, teachers will use these investigations in ways that make sense for their particular style, the particular group of students, and the constraints and supports of a particular school environment. The Teacher Notes and Dialogue Boxes can't provide answers to all of the teachers' questions, but they do provide information and guidance for a wide variety of situations, drawn from collaborations with many teachers and students over many years.

TEACHING MATHEMATICS TO *ALL* STUDENTS

Everybody Counts (National Research Council, 1989) is an important book about ways of making sure *all* students—girls and boys, students from different cultures, students with diverse learning styles, students of different language groups— become more connected with mathematics. The *Investigations* curriculum takes the message of this book very seriously. The curriculum was designed so every student can find himself or herself reflected in it at various places. *Investigations* appeals to the varied interests different students have, whether they involve building cities, taking surveys, plotting class journeys on maps, designing toys, or studying the patterns in a sequence of numbers. As students engage in the investigations, they will sometimes find their own passions fully reflected, and at other times will discover new interests.

Investigating Real-Life Contexts

The conscious and explicit concern with equity with which *Investigations* was designed will be largely invisible to the student. What students will see is a range of activities that appeal to them or are about themselves. The investigations often involve exploring neighborhoods, finding out more about families, and connecting what is being learned at school with the realities of students' lives at home. Students experience pride in themselves and their families when they share data brought from home or report on the strategies their families used when they played number games with them. By using "real stuff" from the community, whether it be menus from local restaurants or maps of the neighborhood, the investigations strengthen students' sense of being included in mathematical study.

Some of the investigations are concerned with the broader world rather than with students' immediate environment. While it is important for students to have a secure sense of themselves, their families, and their communities, they can also use mathematics to expand what they know about the world. Data-gathering investigations are ideal for this purpose: students discover that their classmates come from different countries, have different kinds of family structures (living with extended families, for example), or speak other languages at home. Other investigations lead students to explore the length of the school year or the extent of the use of the metric system among different countries.

The investigations often provide great opportunities for students from a variety of cultures to share their backgrounds. For example, in one field-trial classroom, a very quiet third grader became quite animated when telling the class about the long school year and the intensity of the school day in his native land, Japan.

Supporting Limited-English-Proficiency Students

Language-minority students pose a special challenge in the classroom, particularly when several different language groups are represented. How do teachers help these students connect what they understand—both mathematically and culturally— to the culture of American classrooms? A major way *Investigations* supports these connections is through built-in opportunities, like the ones described above, for students to explore and to learn more about each other's background. Small-group and pair interactions are another vital way to support these students. As students interact, using concrete materials to model problems, all students connect what is being *done* to solve the problem with what is being *said* about solving the problem. Specific curriculum components support teachers' work with limited-English-proficient students:

- **Preview for the Linguistically Diverse Classroom**. Because some working vocabulary must inevitably be assumed by any curriculum material, words that may be unfamiliar to some students have been highlighted in this preview. In some units, this section also includes Multicultural Extensions, which suggest ways students can connect aspects of their daily lives, cultures, and backgrounds to the mathematics content of the unit. Translations of these previews are available in the *Spanish Vocabulary Booklet,* available at each grade level as part of the *Spanish Vocabulary Package.*

- **Vocabulary Support for Second-Language Learners.** This section provides specific suggestions for activities that will help teachers familiarize students with the working vocabulary of the unit in an active, meaningful context. Translations of these sections are in the *Spanish Vocabulary Booklet.*

- **Tips for the Linguistically Diverse Classroom.** Throughout each unit are tips about specific teaching strategies to encourage the participation of all students.

- **Family Letters.** A letter in each unit provides an important means of communicating with student's families. These letters give family members a sense of the mathematical content of the unit, describe how these mathematical ideas relate to aspects of students' daily lives, and suggest ways in which families can support students' mathematics work. These letters are also available separately in six languages.

- **Spanish Blackline Masters.** The *Spanish Vocabulary Package,* available at each grade level, contains Spanish versions of all blackline masters to help teach the *Investigations* curriculum to Spanish-speaking students.

Investigating Mathematical Contexts

Not all mathematics activity in the elementary classroom is about real-life problems or culturally diverse contexts. Some rich and exciting mathematics is about mathematical objects—numbers, geometric shapes, patterns—and their relationships. That is, many important mathematical experiences are *about mathematics itself.* How can these experiences be designed to work with students who bring very different strengths, needs, and previous experiences in mathematics to the classroom? At least three significant aspects of these investigations make them accessible to all students:

- Students can use strategies and tools that are appropriate for them. For example, to figure out the difference between $0.76 and $2.01, students can use counters to count one by one, use coins to model the problem, or use elegant mental strategies. By working with and observing others, students have an opportunity to see a range of strategies and, with the teacher's guidance, try out strategies that are comfortable next steps for them.

- The investigations are designed so students can do them at many different levels. For example, the game Fraction Track can be played by students who are just beginning to understand that $\frac{6}{5} = 1 + \frac{1}{5}$, as well as by students, and even adults, who are using much more complicated numerical relationships, such as $\frac{6}{5} = \frac{2}{10} + \frac{2}{3} + \frac{2}{6}$. Students enter the mathematics at a comfortable level, then quickly find that there are many complexities to explore.

- Speed and memorization are not emphasized. No serious study of any discipline, including mathematics, is about speed. A focus on getting a right answer quickly undermines the value of thinking deeply about the whole problem. Even worse, such a focus becomes a barrier that prevents many students who are not quick at memorizing from achieving success in mathematics. Slowing the pace so students work on a few complex problems carefully is critical both for changing students' view of the nature of mathematical study and for including all students. Quick recall of useful mathematical relationships—such as the addition combinations or knowledge about what happens when any number is multiplied by 10—is developed through students' own construction of knowledge they understand and can rely on, rather than by the rote memorization of a string of symbols.

Mathematics in the elementary grades includes problems that involve whole numbers and fractions, measuring and weighing, collecting data, graphing, spatial visualization, coins and calendars, symmetry and geometric motions. Too often, a narrow mathematical curriculum has failed to uncover the talents and interests of many students. By including significant work in geometry and data, in number theory and the mathematics of change, along with the more traditional content of the number system and number operations, the *Investigations* curriculum ensures that the budding student of geometry—who can "just see" where the puzzle piece fits when no one else can—and the statistician or the pattern finder will also shine in mathematics.

GETTING STARTED WITH THE *INVESTIGATIONS* CURRICULUM

Making Choices for Your School or System

The units in each grade level offer more than enough material for a complete mathematics curriculum. A "standard sequence" of units is suggested (see pp. 4–6), based on experience from classroom testing about which units build on others and which are best done earlier or later in the year. However, there are many alternative ways to arrange the units; the standard sequence gives a place to start, but as schools and teachers become more familiar with the curriculum, they may want to develop a sequence and pacing that makes sense for their situations.

Finding the best pace and sequence may involve changes such as reversing the order of two units, expanding a three-week unit to five weeks to accommodate the need for more extended work on a topic, or spending more time on a particular unit because it relates to a central subject emphasis for the year. For example, one school might decide to emphasize geometry across all grades during a particular year; during that year, classes might spend four to five weeks on a three-week geometry unit, pursuing extensions suggested by the curriculum or created by teachers and students, and spending additional time on interaction and sharing with other grade levels.

Spending more time on some units, of course, necessitates spending less time on others and perhaps omitting some altogether. These are choices schools will have to make, while keeping in mind the following principles:

- Students should encounter significant work in number, data, *and* geometry during each year of the elementary grades.
- There is an abundance of exciting, age-appropriate mathematics in the curriculum to be studied—much more than a class has time to explore in any one school year.
- Students do not master important mathematical ideas during two to four weeks of work, but continue to develop them as they encounter them year after year.
- Units are designed as an integrated set of investigations that allow students to encounter central mathematical ideas in depth and in a variety of contexts. Pulling separate activities out of a unit is not an effective way to use the curriculum.
- However, students do not need to do *every* activity within a unit; extent of coverage is not as important as what is done with what is covered.

Within these principles, variation from year to year (or every few years) may be helpful. There is no single perfect mathematics curriculum for all students—or all teachers. Teachers need not make the same choices every year. They participate actively in deciding which units require more time or less time for a particular group of students. The kind of variation suggested here provides flexibility in responding to differences between last year's class and this year's class and also gives *teachers* a chance for change—the opportunity to stretch themselves intellectually in new ways every two or three years.

Replacement Unit Strategy

A change of curriculum usually means teachers must plunge right into completely new materials and activities for the entire year. However, with the *Investigations* curriculum's modular structure, schools and teachers can choose to implement the curriculum slowly by choosing a few modules—even just one or two at a grade level—to use as replacement units during the first year. The next year, teachers can repeat these units, with which they will now be more comfortable, and add several new units. Some schools may choose to implement fewer units each year than the total available for each grade level, or may choose a core of standard units and vary the others. For example, each grade might choose a core of five or six units that they use each year, while teachers choose two or three from the other available units each year, thus providing flexibility and choice.

If the *Investigations* curriculum is implemented in this way, the introductory unit *(Mathematical Thinking)* at each grade level should be one of the units chosen. These introductory units are designed to support teachers and students as they become familiar with a new set of expectations and are a good way to set the tone for the year's mathematics work. They provide experiences that gradually introduce students to materials, to discussing mathematics, to explaining and writing about their thinking, and to working in pairs and small groups. At the same time, the introductory units provide guidelines and activities for assessing students' mathematical understanding to help teachers plan how to guide and support each student.

When students come from lower grades familiar with the *Investigations* curriculum, teachers may choose to omit the introductory unit in grades 3 and 4 to make more time for exploring other units in depth. At these grades, the introductory units provide only a brief overview and initial assessment; students are not expected to master the ideas encountered. The grade 5 introductory unit contains core material about landmarks in the number system and so should not be omitted.

Grade Levels in the *Investigations* Curriculum

Although each unit is part of the standard sequence for a particular grade level, all units are designated with a grade range. For example, the unit *Money, Miles, and Large Numbers* is part of the grade 4 standard sequence, but is designated as appropriate for grades 4 and 5. As these units were tested in a range of classrooms, very large within-grade variations among students were revealed in terms of their previous mathematics experience, their knowledge of mathematics, their approaches to solving problems, their communication skills, their interest in mathematics, and their confidence as mathematics learners.

This variation is due to the interaction of many factors, from differences in age-level requirements among school systems to unequal opportunities to engage in significant mathematical thinking. Given this wide variation, schools and teachers are encouraged to select from the *Investigations* units and to develop the grade-level sequence that makes most sense for their own situations. For example, because of her students' weak grasp of the structure of the number system, one fourth-grade teacher used both *Landmarks in the Hundreds* (grade 3) and *Landmarks in the Thousands* (grade 4) during a single year.

An entire grade-level sequence of units is also suitable for the next highest grade level. At first, for example, some schools may choose to use the standard grade 3 sequence at grade 4 (or the standard grade 4 sequence at grade 5) because it provides their students with opportunities for significant mathematical thinking about mathematical content they need to learn. Fifth-grade classes that are introducing students to the *Investigations* curriculum for the first time might begin with units on a range of topics from the grade 4 sequence; for example, *Landmarks in the Thousands, The Shape of the Data, Seeing Solids and Silhouettes, Arrays and Shares,* and *Different Shapes, Equal Pieces.*

Once the *Investigations* curriculum has been implemented from the early grades and students have become accustomed to thinking mathematically, those schools may find that they want to start using the standard grade-level sequence.

THE *INVESTIGATIONS* CURRICULUM AND THE NCTM STANDARDS

This overview describes how the *Investigations* units for grades 3–5 develop the mathematical content suggested by the National Council of Teachers of Mathematics' *Curriculum and Evaluation Standards for School Mathematics.* The first four NCTM Standards are discussed below. They are woven throughout the units and must be considered together.

Following the section about the first four standards are three kinds of charts that help to show how the NCTM Standards 5–13 for grades K–4—and, for the grade 5 unit, those for grades 5–8—are addressed by the *Investigations* curriculum:

- **Correlation charts.** For each grade, these charts indicate which of the curriculum units at that grade level address which of the NCTM Standards 5–13.
- **Unit topic charts.** For each grade, these charts list the topics addressed in each unit.
- **Unit-by-unit correlation charts.** For each curriculum unit, this chart lists the key Mathematical Emphases from that unit and indicates which of the NCTM Standards 5–13 are supported by each. Investigation and session numbers where that particular emphasis is first introduced are listed (e.g., 2:3–4 means Investigation 2, Sessions 3 and 4). For example, in the chart for the unit *Introduction: Mathematical Thinking at Grade 3,* the first Mathematical Emphasis— Counting and grouping quantities to make 100— connects to NCTM Standards 5, 6, 9, and 13 and is first introduced in Investigation 1, Session 1 of that unit. The Investigation and session references (e.g., 1:1) will help the user of this guide find examples of how a particular Mathematical Emphasis is incorporated into instruction. However, many of the mathematical understandings and processes described in the Mathematical Emphases are difficult and complex. They are not "learned" in a single investigation but continue to be explored in other investigations, in other units, and often over many years of schooling.

Standard 1
Mathematics as Problem Solving

Problem solving is the heart of the *Investigations* curriculum. In each investigation, students consider problems, develop a variety of strategies to solve them, and share their solutions.

Problems range from small, well-defined problems—There are 12 pencils in a package; how many packages do we need to open so everyone in the class gets one pencil?—to project-sized problems that may take several class sessions: How much taller is a fourth grader than a first grader? In the course of the year, students develop their own approaches to problems, write their own problems, and bring their own creativity and perspective to open-ended problems: Make a 3-D model and determine its volume.

Some problems relate to students' experiences at home, in school, and in the community: How many more days are we out of school than we are in school? How far is it from my home to the school? Can you divide $1.00 evenly among three people? Other problems, just as fascinating to elementary school students, are about mathematical objects and relationships: Write several division problems that have an answer of 0.75. Write a procedure to draw rectangles similar to a 10 by 15 rectangle. Which of the tetrominoes will perfectly cover a 10 by 12 rectangle? What's the shortest path between two points on a coordinate grid? The availability of a range of tools—manipulative materials, calculators, and computers—gives students more ways to approach problems and allows them to demonstrate and explain their solutions.

In all of these problem-solving activities, students are considering questions *about problem solving:* How do I know when I've solved the problem? Am I sure of my solution? How can I double-check it? Are my conclusions based on the data? Is this always true? Which of these are impossible? By providing investigations that encourage students to think mathematically in these ways, the *Investigations* curriculum offers a problem-centered approach to mathematics in the elementary classroom.

Standard 2
Mathematics as Communication

In each *Investigations* unit, students are involved in building, drawing, representing, writing, and talking as part of their mathematics work. Students develop their own strategies for representing and recording and are introduced to a repertoire of useful ways of using concrete materials, tables, graphs, and charts. For example, in *Containers and Cubes* (grade 5), students use interlocking cubes to model the volume of a box. In *Changes Over Time* (grade 4), students develop representations to show who enters and leaves their homes over a 24-hour period; they also learn to use line graphs to keep track of the growth of their bean plants.

Reflection and discussion are key to the mathematical activity in *Investigations*. Throughout each unit, students work and talk together in pairs and small groups, then periodically come together as a whole class to discuss their work. The importance of these discussion times is emphasized by the inclusion of Dialogue Boxes in the teacher books, which provide examples of ways in which students may talk about mathematical ideas and of the kinds of confusion with which students are likely to struggle.

The embedded assessment items include writing, drawing, and representing activities. Different students express themselves best in different modes; therefore, assessment activities include various ways for students to demonstrate their understanding. In some assessment activities, students construct a solution using concrete materials; by walking around and scanning students' constructions, teachers can quickly get a sense of how students are understanding the mathematical ideas. At least one assessment in each unit involves writing or drawing, so teachers can keep track through the year of students' developing ability to show a solution on paper. For example, in one assessment activity in *Arrays and Shares* (grade 4), students solve the problem 27×4 in two different ways, then write about their solution strategies.

Because of the emphasis on communication, student sheets on which students record their work are not always included for every activity. Instead, in some situations, students are required to develop their own form and organization for recording their work so that they begin to develop skill and confidence in keeping track of their work and communicating their thinking to others.

To support the emphasis on communicating mathematically, the units include examples of classroom discussions and samples of student work. For example, in the unit *Seeing Solids and Silhouettes* (grade 4), several examples of students' instructions for making a toy from interconnecting cubes

are shown and discussed to help teachers assess the mathematical skills and ideas students are using in their work. Teacher Notes also provide support for teachers as they help students represent, record, and discuss their work.

Standard 3
Mathematics as Reasoning

The *Investigations* curriculum is designed to support teachers and students as they move away from a view of mathematics as a series of facts and procedures to be memorized toward a view of mathematics as a discipline in which one can use all the resources at hand to reason about mathematical problems. Students learn to solve a problem by thinking about what they already know, making pictures and models, trying examples, and listening to their peers. Mathematics becomes a science in which a situation is observed, conjectures are made, evidence is considered, and conclusions are based on analyzing the data.

In this view, mathematics is not a series of individual problems. Each problem is related to a web of other problem situations. Students are encouraged to use previous knowledge to solve new problems, and to reflect on solutions to new problems as a way of expanding their knowledge of mathematical relationships. The teachers' role is critical. Rather than simply accepting correct answers, teachers push students' thinking further by encouraging them to think hard about mathematical relationships. For example, during the unit *Up and Down the Number Line* (grade 3), students in one classroom were working on relationships among positive and negative numbers. By looking at many examples, most students had decided that the addition of opposite numbers (such as +2 and −2) results in zero. However, the teacher pushed students to think harder about why this relationship is true—would it always be true no matter what numbers were considered?

The emphasis on mathematical reasoning must pervade all of the elementary mathematics content—including the development of strategies for the basic arithmetic operations. Students develop and apply sound number sense only when they are encouraged to build on what they know to reason about numerical problems. For example, in *Landmarks in the Thousands* (grade 4), students learn about important landmarks in the number system—such as multiples of 10 and 100—that they can use to reason about numerical problems. How many 20's are in 940? A student reasons from her knowledge that there are five 20's in every 100, so in 9 hundreds, there must be 45 twenties. There are two more 20's in 40, for a total of 47 twenties in 940. What might be considered a hard division problem is solved easily when students are encouraged to develop a rich repertoire of knowledge about number relationships and to use this knowledge to reason about new problems.

When students know their teachers are genuinely interested in how they think—not just in whether their answers are correct—they gradually learn to share even tentative ideas and to explore the ideas of other students. In all the *Investigations* units, students are asked to think hard about mathematics—to reason from what they know, from what they can construct or model, and from evidence they can gather to solve mathematical problems.

Standard 4
Mathematical Connections

The *Investigations* curriculum supports making connections among ideas within mathematics and between mathematical ideas and the many contexts to which they are connected.

Most units are structured around connected mathematical ideas. Multiplication and division always appear together, as do addition and subtraction. These operations are not viewed as separate parts of mathematics: problems and solution strategies are closely intertwined. In fact, many people use addition to solve subtraction problems and multiplication to solve division problems. Making such connections is explicitly supported in the *Investigations* materials. Many other connections of mathematical content appear in the way units are structured. In several units, fraction ideas are used to interpret data. In some units, rectangular arrays are used to represent numbers and their factors. In *Flips, Turns, and Area* (grade 3), geometric motions and area concepts are treated together as students explore the geometric shapes known as *tetrominoes*.

Throughout the units, the areas of data, number, and geometry are carefully integrated. Geometric models are used in many number units. Pattern blocks provide an important model in *Fair Shares* (grade 3), while making rectangles from tiles becomes a way to explore factors and multiples in *Landmarks in the Thousands* (grade 4). Counting, measuring, and other number work support data collection and analysis. Data sets collected by students—for example, in *Combining and Comparing* (grade 3)—are an important context for developing strategies for addition and subtraction.

Perhaps the most important connections made in this curriculum are between conventional mathematical symbols, terms, and notation and the *meaning* of mathematical operations and relations. The units emphasize the care that must be taken when introducing conventional notation and mathematical terms so these terms and notation support students in recording and describing what has been demonstrated, visualized, represented, and understood. Mathematical vocabulary is introduced naturally as it helps students articulate ideas they are thinking about, and notation is presented as a conventional way of recording relationships that students understand (see, for example, the Teacher Notes, Introducing Mathematical Vocabulary and What About Notation? that appear in several units, including *Arrays and Shares,* grade 4).

It is critical that students view mathematics not as an isolated discipline but as a way of thinking that connects to aspects of daily life as well as to other disciplines such as science, social studies, and literature. The variety of familiar contexts used throughout the *Investigations* units—sharing, ages, toys, heights, pets, package labels, giving directions—supports the connection between students' experiences in their lives and the mathematics work they do at school. Science (for example, investigating the change of weight in pieces of fruit as they dry, growing and recording the height of bean plants), social studies (studying changes in the population of familiar places in the community, planning a trip within the United States to investigate playground accidents), and children's literature (the use of *How Big is a Foot?* by Rolf Myller in *From Paces to Feet,* grade 3) all provide rich contexts for mathematical investigations.

GRADE 3 OVERVIEW OF CORRELATION TO NCTM STANDARDS

Grade 3 Units	Estimation	Number Sense and Numeration	Whole Number Operations	Whole Number Computation	Geometry and Spatial Sense	Measurement	Statistics and Probability	Fractions and Decimals	Patterns and Relationships
	5	6	7	8	9	10	11	12	13
Mathematical Thinking at Grade 3 (Introduction)		✔	✔	✔	✔				✔
Things That Come in Groups (Multiplication and Division)		✔	✔	✔	✔				✔
Flips, Turns, and Area (2–D Geometry)					✔	✔			✔
From Paces to Feet (Measuring and Data)	✔				✔	✔	✔		✔
Landmarks in the Hundreds (The Number System)		✔	✔	✔	✔				✔
Up and Down the Number Line (Changes)		✔	✔	✔			✔		✔
Combining and Comparing (Addition and Subtraction)	✔	✔	✔	✔		✔	✔		✔
Turtle Paths (2-D Geometry)			✔	✔	✔	✔			
Fair Shares (Fractions)					✔			✔	✔
Exploring Solids and Boxes (3-D Geometry)	✔	✔			✔				✔

Grade 3 Unit Topics

Mathematical Thinking at Grade 3 (Introduction)

- Counting and grouping
- Number patterns on the 100 chart
- Odd and even numbers
- Symmetry
- Collecting, recording, and representing data
- Addition and subtraction strategies and combinations
- Combining coins

Things That Come in Groups (Multiplication and Division)

- Multiplication with groups
- Patterns and multiples on a 100 Chart
- Multiplication with arrays
- Relationships between multiplication and division

Flips, Turns, and Area (2–D Geometry)

- Measuring areas by covering spaces with units and half-units
- Comparing areas of rectangles with different dimensions
- Describing motions as slides, flips, and turns
- Comparing shapes for congruence

From Paces to Feet (Measuring and Data)

- Estimating length
- Measuring in nonstandard and standard units
- Using U.S. standard and metric measures
- Collecting, describing, and analyzing data

Landmarks in the Hundreds (The Number System)

- Counting and grouping
- Factors and multiples
- Multiples of 100
- Multiplication and division

Up and Down the Number Line (Changes)

- Numbers above and below zero
- Net change
- Inverse operations—addition and subtraction
- Graphing positive, negative, and zero change

Combining and Comparing (Addition and Subtraction)

- Developing strategies for combining and comparing numbers
- Collecting and comparing data
- Numeration through hundreds and thousands
- Using measurement to combine heights and compare weights
- Adding amounts to make a sum of money
- Exploring mathematical characteristics of the calendar

Turtle Paths (2-D Geometry)

- Paths and lengths of paths
- Turns
- Describing triangles
- Measuring paths, turns, and perimeter

Fair Shares (Fractions)

- Making equal parts
- Grouping unit fractions
- Equivalent fractions
- Mixed numbers
- Relating fractions and division
- Fractions and decimals

Exploring Solids and Boxes (3-D Geometry)

- Sorting and describing solids
- Building polygons
- Building polyhedra
- Investigating boxes and patterns for solids

GRADE 3 UNIT-BY-UNIT CORRELATION TO NCTM STANDARDS

Mathematical Thinking at Grade 3 (Introduction)

Students are introduced to content, processes, and materials for solving problems in mathematics. They are introduced to a way of approaching mathematics that emphasizes thinking, strategy use, communication and collaboration.

Mathematical Emphases	Estimation	Number Sense and Numeration	Whole Number Operations	Whole Number Computation	Geometry and Spatial Sense	Measurement	Statistics and Probability	Fractions and Decimals	Patterns and Relationships
	5	6	7	8	9	10	11	12	13
Counting and grouping quantities to make 100	1:1	1:1			1:1				1:1
Becoming familiar with number patterns on the 100 chart		1:2-3	1:2-3		1:2-3				1:2-3
Exploring materials, including the calculator, to be used throughout the curriculum as tools for solving problems		1:1,2-3 2:1, 5-7 3:3-4 4:2 TMM	1:1,2-3 3:3-4 4:2 TMM	1:1,2-3 3:3-4 4:2	2:1				
Using grouping to count	3:3-4	1:1 3:3-4							
Constructing symmetrical patterns		2:1, 3-4	2:1, 3-4		2:1, 3-4			2:1, 3-4	2:1
Learning the addition combinations from 1 + 1 to 10 + 10		2:1, 2, 3-4	2:1, 2, 3-4 TMM	2:1, 2, 3-4					2:1, 2
Developing and using strategies to combine and compare quantities	3:3-4	2:2, 5-7 3:3-4 4:1 TMM	2:2, 5-7 3:3-4 4:1 TMM	2:2, 5-7 3:3-4 4:1 TMM					2:5-7 4:1
Exploring what happens when you add or subtract 10 or 20		1:1 2:3-4, 5-7	1:1 2:3-4, 5-7	1:1 2:3-4, 5-7					1:1 2:3-4, 5-7
Exploring what numbers can be divided evenly		2:3-4 4:2	2:3-4						
Reviewing the values of coins and finding the values of collections of coins		2:5-7 TMM	2:5-7 TMM	2:5-7 TMM					
Sorting and classifying information		3:3-4					3:1-2 3:3-4		3:1-2
Collecting, recording, and representing data		3:3-4					3:1-2, 3-4 TMM		3:1-2, 3-4 TMM
Exploring the characteristics of odd and even numbers and how they behave when combined		4:1,2,3	4:1,3	4:1					4:1
Working with wholes and halves		4:2	4:2	4:2				4:2	4:2
Developing awareness of the decimal point and its meaning		4:2	4:2					4:2	

TMM = Ten-Minute Math
1:2–3 = Investigation 1, Sessions 2–3

Things That Come in Groups (Multiplication and Division)

Students work with things that come in groups, with patterns in the multiplication tables using 100 charts, and with rectangular arrays. They invent and solve problems in multiplication and division.

	NCTM Standards								
Mathematical Emphases	Estimation	Number Sense and Numeration	Whole Number Operations	Whole Number Computation	Geometry and Spatial Sense	Measurement	Statistics and Probability	Fractions and Decimals	Patterns and Relationships
	5	6	7	8	9	10	11	12	13
Finding things that come in groups		1:1, 2	1:2	1:2					
Using multiplication to mean groups		1:1,2 2:3-4	1:1,2, 3,4 2:3-4	1:2,3,4 2:3-4					1:1,2
Recognizing that skip counting represents multiples of the same number and has a connection to multiplication		2:1,2, 3-4 5:3 TMM	1:4 2:1,2, 3-4 5:1 TMM	5:1					TMM 2:2
Finding patterns in multiples of 2, 3, 4, 5, 6, 9, 10, 11, and 12 by using the 100 chart and the calculator		2:1,2, 3-4,5-6	2:1,2, 3-4,5-6						2:1,2, 3-4,5-6
Understanding that number patterns can help in multiplication			2:3-4	2:3-4					2:3-4
Recognizing that multiplication can be used to find the area of a rectangle		3:3	3:1,2,3	3:3	3:1,2,3				
Using arrays to skip count; multiplying and dividing with skip counting		3:2,3	3:1,2,3	3:2,3	3:1,2,3				3:2
Finding factor pairs			3:2,3	3:2,3 4:1	3:2				3:3 4:1
Understanding relationships between multiplication and division			1:3 3:3 4:1 5:4	3:3 4:1 5:4	3:3				3:3 5:4
Identifying whether word problems can be solved using division and/or multiplication			4:1,3-4	4:1,3-4					
Using multiplication and/or division notation to write number sentences			1:2,4 2:3-4 4:1	4:1					
Using patterns to solve multiplication and division problems			3:3	3:3 5:1	3:3				3:3 5:1
Organizing and presenting data in tables and line plots			5:1,3	5:1,3				5:1,3	5:1
Sorting out complex problems that require both multiplication and addition		5:4	5:1,4	5:1,4				5:1	5:4
TMM—Describing events as likely and unlikely							4:2 5:1 TMM		

TMM = Ten-Minute Math
1:2–3 = Investigation 1, Sessions 2–3

Flips, Turns, and Area
(2-D Geometry)

Students develop spatial visualization abilities as they investigate, measure, and compare area of shapes. They explore geometric motions—slides, flips, and turns—as well as measuring area in units and half-units.

Mathematical Emphases	Estimation	Number Sense and Numeration	Whole Number Operations	Whole Number Computation	Geometry and Spatial Sense	Measurement	Statistics and Probability	Fractions and Decimals	Patterns and Relationships
	5	6	7	8	9	10	11	12	13
Measuring area by covering a flat space with square units		1:2-3 2:2-3, 4-5	1:2-3, 4-5	2:4-5	1:1,2-3 2:1, 2-3,4-5	1:1,2-3 2:1, 2-3,4-5			1:2-3
Systematically finding all possible geometric arrangements of a given number of squares					1:1				1:1
Finding patterns for covering a space					1:1,2-3	1:1,2-3			1:1,2-3
Comparing areas of rectangles that have different dimensions		1:4	1:4,5		1:4,5	1:4,5			1:4,5
Describing physical motions precisely as a series of slides, flips, and turns					1:1, 2-3,5 2:2-3				1:2-3
Comparing the area of two shapes by determining whether they cover the same amount of flat space					2:2-3	2:2-3			
Comparing shapes to determine congruence through motions such as rotation (turns) and reflection (flips)					2:2-3, 4-5				2:2-3
Exploring relationships among shapes (for example, a rectangle can be cut into two triangles, each of which is half the area of the rectangle)					2:2-3, 4-5	2:2-3, 4-5		2:2-3	
Finding the area of complex shapes by identifying smaller units of area, such as square units and half units					2:1,4-5	2:1,4-5			
TMM—Finding alternative ways to arrive at the same numerical solution		1:2-3 2:2-3 TMM	1:2-3 2:2-3 TMM	1:2-3 2:2-3 TMM					1:2-3 2:2-3 TMM

TMM = Ten-Minute Math
1:2–3 = Investigation 1, Sessions 2–3

From Paces to Feet (Measuring and Data)

Students explore the need for standard measurement, learn to use different measuring tools and systems, and interpret data they collect by measuring.

Mathematical Emphases	Estimation 5	Number Sense and Numeration 6	Whole Number Operations 7	Whole Number Computation 8	Geometry and Spatial Sense 9	Measurement 10	Statistics and Probability 11	Fractions and Decimals 12	Patterns and Relationships 13
Using a nonstandard unit to measure a distance and experiencing the iterative nature of measurement	1:1, 2, 3-4				1:1, 2, 3-4	1:1, 2, 3-4	1:1, 2		
Estimating length in "paces" by visualizing the unit "pace" repeated over a distance	1:1, 2, 3-4				1:1, 2, 3-4	1:1, 2, 3-4			
Comparing the effects of measurement using units of different sizes	1:2				1:1, 2	1:1, 2 2:2	1:1, 2 2:2		2:2
Describing the shape of the data and analyzing it for patterns							1:2		1:2
Examining a set of data to determine which is the "middle-sized" piece					1:5-6	1:5-6 2:2	1:5-6 2:2		1:5-6 2:2
Understanding the rationale for a standard measure					1:5-6 2:1	1:5-6 2:1			
Developing familiarity with inches, feet, and yards					2:1, 3-4	2:1, 2, 3-4	2:1, 3-4		2:1, 2
Developing awareness of centimeters and meters and how big these units of measure are					2:5, 6-7 4:1-3	2:5, 6-7 4:1-3	2:6-7		
Describing a set of data that involve measurements by representing the data on a line plot and then by describing the general features of the data					2:3-4	2:2, 3-4, 5, 6-7	2:2, 3-4, 5, 6-7		2:2, 3-4, 5
Using standard measures (U.S. standard or metric) in complex situations to gather and analyze data concerning size and proportion					3:2-3 4:1-3	2:6-7 3:1,2-3 4:1-3	2:6-7 3:2-3		3:2-3
TMM—Estimating solutions to arithmetic problems and using mental computation strategies to find an answer	1:2, 5-6 TMM	1:2, 5-6 TMM	1:2, 5-6 TMM	1:2, 5-6 TMM					
TMM—Developing a visual image of a geometric figure					2:2, 5 TMM				

TMM = Ten-Minute Math
1:2–3 = Investigation 1, Sessions 2–3

Landmarks in the Hundreds (The Number System)

Students work with 100, investigating factors of 100, and multiples of 100 (up to 1000). Based on their understanding of landmark numbers, they develop strategies to solve multiplication and division problems.

Mathematical Emphases	Estimation	Number Sense and Numeration	Whole Number Operations	Whole Number Computation	Geometry and Spatial Sense	Measurement	Statistics and Probability	Fractions and Decimals	Patterns and Relationships
	5	6	7	8	9	10	11	12	13
Understanding the relationship between skip counting and grouping		1:1,2-3	1:2-3		1:1,2-3				1:1,2-3
Becoming familiar with the relationships among commonly used factors and multiples		1:1,2-3	1:1,2-3		1:1,2-3				1:1,2-3
Increasing fluency in counting by single-digit numbers and by useful two-digit numbers		1:1,2-3 2:1-3 TMM							1:2-3 TMM
Developing familiarity with the factors of 100 and their relationships to 100 using cubes, coins, and 100 charts		1:3-4, 6-7 TMM	1:3-4, 6-7 TMM		1:3-4			1:6-7	1:3-4, 6-7 TMM
Using knowledge about factors of 100 to understand the structure of multiples of 100		2:1-3	2:1-3	2:1-3					2:1-3
Developing strategies to solve problems in multiplication and division situations by using knowledge of factors and multiples		1:6-7 2:5-6	1:6-7 2:4,5-6	2:1-3, 5-6					2:1-3
Reading and using standard multiplication and division notation to record			1:6-7 2:5-6	2:5-6					
Using factors of 100 to understand the structure of 1000		3:1 TMM	3:1,2-3		3:2-3				3:2-3
Estimating quantities up to 1000	2:5-6 3:2-3	3:2-3							
Using landmarks to calculate "distances" within 1000 (How far is it from 650 to 950?)		3:2-3	3:2-3						
Creating numerical expressions that equal a given number		TMM	TMM	1:6-7 TMM					

TMM = Ten-Minute Math
1:2-3 = Investigation 1, Sessions 2-3

Up and Down the Number Line (Changes)

Students investigate addition and subtraction as they work with movement on the vertical and then horizontal number lines. They explore numbers below and above zero, create graphs showing positive, negative, and zero change, and identify net change.

Mathematical Emphases	Estimation	Number Sense and Numeration	Whole Number Operations	Whole Number Computation	Geometry and Spatial Sense	Measurement	Statistics and Probability	Fractions and Decimals	Patterns and Relationships
	5	6	7	8	9	10	11	12	13
Finding net (total) change given a starting and ending number		1:1, 2	1:1,2, 3,4	1:1,2, 3,4	1:3,4				1:1, 2
Using subtraction to cancel addition		1:5, 8	1:3,4, 5,8	1:3,4,5					1:3,4,5
Making the same net change in many different ways using positive and negative numbers			1:3,4, 6,7	1:3,4, 6,7					
Using net change to determine an end point instead of counting each change separately		1:5 3:1, 2	1:5 3:1, 2	1:5 3:1, 2					
Developing strategies for adding a long sequence of changes, including number and operation sense; using a calculator	TMM	TMM 1:3,4,5	TMM 1:5	TMM 1:3,4,5					
Developing strategies for finding a missing starting number or a previous position along the number line		1:6-7	1:6-7	1:6-7					1:6-7
Representing numbers graphically and understanding that a "going up" graph indicates positive change, a "going down" graph indicates negative change, and a horizontal graph indicates zero change		2:1,2,3	2:1,2,3				2:1,2,3		2:2,3
Finding net change on graphs			2:2,3				2:2,3		
Recognizing that passage of time or order of events can be represented by moving from left to right			2:2,3				2:2,3		2:2,3,4
Moving to the left for negative changes and to the right for positive changes		2:4 3:1,2	2:4 3:1,2	3:1,2			2:4		2:4
Halving and doubling numbers				3:1					

TMM = Ten-Minute Math
1:2–3 = Investigation 1, Sessions 2–3

Grade 3 Correlation to NCTM Standards 5–13 ■ 29

Combining and Comparing (Addition and Subtraction)

Students solve problems that involve comparison of quantities, data, or measurements. They are encouraged to develop their own addition and subtraction strategies, to estimate, and to use multiple strategies to double-check their work. They build fluency with using numerical landmarks in combining and comparing.

Mathematical Emphases	Estimation 5	Number Sense and Numeration 6	Whole Number Operations 7	Whole Number Computation 8	Geometry and Spatial Sense 9	Measurement 10	Statistics and Probability 11	Fractions and Decimals 12	Patterns and Relationships 13
Developing computation strategies for combining and comparing based on number sense and number relationships	2:2	1:1,2 2:2 3:1-2 4:3-4 TMM	1:1,2 2:2 3:1-2 4:3-4 TMM	1:1,2 2:2 3:1-2 4:3-4 TMM		3:1-2	1:1,2		3:1-2
Using landmark numbers (multiples of 10 and 100) in comparing and combining quantities	1:1 2:1,2 3:1	1:1,2 2:1,2 3:1 4:2,3-4 TMM	1:1,2 2:1,2 3:1 4:2,3-4 TMM	1:1,2 2:1,2 3:1 4:2,3-4 TMM					4:3-4
Examining how parts and the whole are related in addition and subtraction			5: 2-3						5: 2-3
Solving addition problems with multiple addends	3:1-2 TMM	3:1-2,3 TMM	3:1-2,3 TMM	3:1-2,3 TMM		3:1-2,3	3:1		
Developing more than one way to solve a computation problem and using one method to check another	3:1	3:1 5: 2-3	3:1 4:1 5: 2-3 TMM	3:1 4:1 5: 2-3 TMM		5: 3-4	5: 3-4		
Solving compare and combine problems with strategies and recording with standard addition and subtraction notation	1:1 4:2	1:1 4:2	1:1 4:2	1:1 4:2					
Making comparisons of how things change over time	2:1	2:2	2:2	2:2		2:1,2	2:1,2		
Learning to weigh with a pan balance						2:1,2	2:1,2		
Exploring number relationships in the context of time, money, and linear measure	3:2	3:2 5: 1,2-3	3:2 5: 1,2-3	3:2 5: 1,2-3	3:2	5: 1,2-3			3:2
Using important equivalencies of time, money, and linear measure		3:2	3:2	3:2	3:2,3				3:2
Estimating solutions that can be adjusted to construct an exact solution	TMM	TMM	TMM	TMM					
Reading and writing numbers in the hundreds and thousands	4:3-4	4:3-4							
Developing strategies to combine and compare quantities in the hundreds and thousands	4:1	4:1,2, 3-4	4:1,2, 3-4				4:1		
Developing conjectures and predictions; evaluating data and evidence	2:1 4:1,2	2:2 4:1,2 5: 2-3	2:2 5: 2-3	2:2 5: 2-3		2:2 5: 2-3	1:1,2 2:2 5: 2-3 TMM		
Collecting, recording, and graphing data		4:1 TMM	4:1				5: 2-3 TMM		
Describing and interpreting data		2:2 4:1 TMM	4:1,2	4:2		2:2	4:1,2 5: 2-3 TMM		
Exploring the mathematical characteristics of the calendar		5:1	5:1	5:1		5:1	5:1		5:1
Developing strategies for problems that combine addition and subtraction		TMM	TMM	TMM					

30 ■ *Grade 3 Correlation to NCTM Standards 5–13*

TMM = Ten-Minute Math
1:2–3 = Investigation 1, Sessions 2–3

Turtle Paths (2-D Geometry)

Students explore problems involving paths, lengths of paths, perimeter, and turns. They do computer activities, using the program *Geo-Logo*™, as well as noncomputer activities to investigate these topics.

Mathematical Emphases	Estimation	Number Sense and Numeration	Whole Number Operations	Whole Number Computation	Geometry and Spatial Sense	Measurement	Statistics and Probability	Fractions and Decimals	Patterns and Relationships
	5	**6**	**7**	**8**	**9**	**10**	**11**	**12**	**13**
Understanding paths as representations or records of movement		1:3-4	1:1, 3-4	1:1	1:1, 2, 3-4 2:1-2	1:1, 2, 3-4 2:1-2			
Finding different ways to meet geometric constraints		1:1	1:1	1:1	1:1,3-4	1:1,3-4			
Using Logo commands to construct paths and describe the properties of paths					1:1,2,3 2:2	1:1,2,3 2:2			
Applying mathematical processes such as addition, subtraction, estimation, and "undoing" to paths in solving geometric problems	2:1-2	TMM 2:1-2	1:2, 3-4 TMM	1:2,3 TMM	1:2,3-4 2:1-2	1:2,3-4 2:1-2 TMM			1:3-4
Understanding turns as a change in orientation or heading		2:1-2			1:1,3-4 2:1-2	1:1 2:1-2			
Estimating and measuring turns (creating, using, and iterating units of turn)	2:1-2	2:1-2	2:1-2		2:1-2 2:4	2:1-2 2:4			
Becoming familiar with a common measurement for turns—degrees—and understanding that there are 360° in one full turn, 180° in a half turn, and 90° in a quarter turn	2:1-2	2:1-2			2:1-2	2:1-2			
Building a definition of triangles from examples, and applying the definition to new figures					2:3,4	2:4			2:3
Using Logo commands to draw equilateral triangles, estimating turn measures and using trial and error strategies		2:4			2:3,4	2:4			2:4
Applying mathematical processes, such as quantitative reasoning, mental arithmetic, and logic, to find missing measures of figures		2:5-6 3:Exc	2:5-6 3:Exc	2:5-6 3:Exc	2:5-6 3:Exc	2:5-6 3:Exc			
Constructing geometric figures that satisfy given criteria using analysis of geometric situations, arithmetic, and problem-solving strategies		TMM 3:1-2	TMM 2:5-6 3:1-2, 3-5	2:5-6 3:1-2, 3-5	2:3,4,5 3:1-2, 3,4,5 TMM	2:5 3:1-2, 3,4,5 TMM			TMM
Linking visual paths to Logo commands to describe, analyze, and understand geometric figures		3:1-2 3:6-7	3:1-2 3:6-7	3:6-7	3:1-2 3:6-7	3:1-2 3:6-7			3:1-2
Understanding that shapes can be moved in space without losing their properties					2:4 3:3-5				
Estimating and measuring the perimeter of various objects		3:1-2, 6-7 TMM	3:1-2, 6-7 TMM		3:1-2, 6-7 TMM	3:1-2, 6-7 TMM			3:1-2

TMM = Ten-Minute Math
1:2–3 = Investigation 1, Sessions 2–3

Fair Shares (Fractions)

Students use fractions and mixed numbers as they solve sharing problems and build wholes from fractional parts. They connect fractions to division, and use the calculator to see fractions as decimals.

Mathematical Emphases	Estimation	Number Sense and Numeration	Whole Number Operations	Whole Number Computation	Geometry and Spatial Sense	Measurement	Statistics and Probability	Fractions and Decimals	Patterns and Relationships
	5	6	7	8	9	10	11	12	13
Realizing that fractional parts must be equal (e.g. one third is not just one of three parts but one of three equal parts)					1:1,2 2:7 3:3	1:1, 2 2:7		1:1, 2 2:7	1:1, 2
Developing familiarity with conventional fraction words and notation (though students can write their solutions in any way that communicates accurately; e.g. a student might write $\frac{1}{2} + \frac{1}{4}$ as "half plus another piece that is half of the half")		1:1, 2						1:1,2, 3,4	1:1, 2
Becoming familiar with grouping unit fractions, those that have a numerator of 1 (for example: $\frac{1}{6} + \frac{1}{6} + \frac{1}{6}$ is equivalent to $\frac{3}{6}$)					2:1, 2			1:1, 2 2:1,2,4	
Developing familiarity with common equivalents, especially relationships among halves, thirds, and sixths (for example, students exchange $\frac{2}{6}$ for $\frac{1}{3}$; they may also begin to make exchanges based on $\frac{1}{6} + \frac{1}{3} = \frac{1}{2}$)					2:1-2, 4, 5-6, 7 3:1-2			2:1-2, 4, 5-6, 7	2:1,2
Understanding that the relationships that occur between 0 and 1 also occur between any consecutive whole numbers (e.g. $\frac{1}{2} + \frac{1}{6} = \frac{2}{3}$ so $2\frac{1}{2} = 2\frac{2}{3}$)					2:3, 4			1:3,4 2:4 TMM	
Understanding the relationship between fractions and division (e.g. by solving problems in which the whole is a number of things rather than a single thing, and the fractional part is a group of things as well, as in $\frac{1}{3}$ of 6 is 2)					1:3,4 2: 5, 6	1:3,4		1:3, 4 2:3,5,6 3:1,2,3	2:5, 6
Relating notation for common fractions ($\frac{1}{2}, \frac{1}{4}, \frac{3}{4}, \frac{1}{5}, \frac{1}{10}$) with notation for decimals on the calculator (0.5, 0.25, 0.75, 0.2, 0.1)								3:1,2	3:1,2
Using different notations for the same problem (e.g. 6 ÷ 2 and $\frac{1}{2}$ of 6)				3:3				3:3	
TMM—Using logical reasoning and number sense to identify a number								1:3-4	
TMM—Developing flexibility in solving problems by finding several ways to reach a solution		2:1, 2		2:1, 2				3:1,2,3 TMM	

TMM = Ten-Minute Math
1:2–3 = Investigation 1, Sessions 2–3

Exploring Solids and Boxes
(3-D Geometry)

Students investigate various polygons and geometric solids. They become familiar with the components of these shapes and explore relationships as they sort, build, and make patterns for solids.

Mathematical Emphases	Estimation	Number Sense and Numeration	Whole Number Operations	Whole Number Computation	Geometry and Spatial Sense	Measurement	Statistics and Probability	Fractions and Decimals	Patterns and Relationships
	5	6	7	8	9	10	11	12	13
Exploring, sorting, comparing, and talking about common geometric solids					1:1				1:1
Investigating and analyzing the parts of solids					1:1,2				
Recognizing the components of polygons— the sides, vertices, and angles					2:1,2				2:1,2
Recognizing how the components of polygons are put together to form whole shapes					2:1,2 3:1				2:1,2
Recognizing the components of polyhedra—the faces, corners, and edges		2:3			1:2 2:3				2:3
Recognizing how the components of polyhedra are put together to form whole shapes		2:3,4,5			2:3,4,5				2:3
Exploring two-dimensional geometric patterns that fold up to make three-dimensional shapes					3:1,2				
Investigating interrelationships between parts of solids					2:3,4-5 3:1				
Improving spatial visualization skills					2:1,3, 4-5 3:1,2 4:3 5: 1–4				2:1,3 TMM
Predicting the number of cubes that fit in a box without (and later with) a top by examining a pattern that makes the box	4:1	4:1			4:1				
Determining the number of cubes that fit in a rectangular box	4:1	4:1 5: 1-4			4:1 5: 1-4	4:1 5: 1-4			
Understanding the structure of rectangular prism arrays of cubes					4:1,2,3 5: 1-4	4:1 5: 1-4			4:2 5: 1-4
Designing patterns for boxes that will hold a given number of cubes					3:1,2 4:2	4:2			
Using appropriate computation techniques to determine the total number of cubes in paper cities				5: 1-4	5: 1-4				
TMM—likely or unlikely, based on a sample—categorizing events as likely or unlikely							4:2 5: 1-4		

TMM = Ten-Minute Math
1:2–3 = Investigation 1, Sessions 2–3

GRADE 4 OVERVIEW OF CORRELATION TO NCTM STANDARDS

Grade 4 Units	NCTM Standards								
	Estimation	Number Sense and Numeration	Whole Number Operations	Whole Number Computation	Geometry and Spatial Sense	Measurement	Statistics and Probability	Fractions and Decimals	Patterns and Relationships
	5	6	7	8	9	10	11	12	13
Mathematical Thinking at Grade 4 (Introduction)	✔	✔	✔	✔	✔				✔
Arrays and Shares (Multiplication and Division)		✔	✔	✔	✔				✔
Seeing Solids and Silhouettes (3-D Geometry)					✔				
Landmarks in the Thousands (The Number System)		✔	✔	✔					✔
Different Shapes, Equal Pieces (Fractions and Area)		✔			✔	✔		✔	✔
The Shape of the Data (Statistics)		✔				✔	✔		✔
Money, Miles, and Large Numbers (Addition and Subtraction)	✔	✔	✔	✔		✔		✔	
Changes Over Time (Graphs)		✔				✔	✔		✔
Packages and Groups (Multiplication and Division)		✔	✔	✔					✔
Sunken Ships and Grid Patterns (2-D Geometry)					✔	✔			✔
Three out of Four Like Spaghetti (Data and Fractions)	✔						✔	✔	✔

Grade 4 Unit Topics

Mathematical Thinking at Grade 4 (Introduction)

- Exploring hundreds
- Grouping and ordering strategies
- Addition and subtraction patterns
- Geometric patterns
- Mirror and rotational symmetry

Arrays and Shares (Multiplication and Division)

- Grouping strategies for multiplication
- Array strategies for multiplication
- Multiplication patterns and relationships
- Dividing quantities into equal shares
- Partitioning quantities into equal parts
- Identifying and creating multiplication and division situations

Seeing Solids and Silhouettes (3-D Geometry)

- Identifying and describing spatial relationships
- Developing visualization skills
- Using geometric perspective to draw views of 3-D figures
- Using views to visualize and/or construct 3-D figures

Landmarks in the Thousands (The Number System)

- Exploring factors of 100
- Identifying and using landmarks in the number system to 10,000
- Identifying and using multiples of 100 and 1000
- Using landmarks to add and subtract

Different Shapes, Equal Pieces (Fractions and Area)

- Constructing fractions that are equal in area but are not congruent
- Exploring relationships among fractions: halves, fourths, eighths; thirds, sixths, twelfths
- Combining fractions to make a whole
- Comparing fractions to the landmark numbers $0, \frac{1}{2}, 1,$ and 2
- Ordering fractions
- Identifying equivalent fractions

The Shape of the Data (Statistics)

- Collecting, organizing, and representing data
- Analyzing data sets
- Comparing data sets
- Finding and describing the median
- Carrying out a data-analysis investigation

Money, Miles, and Large Numbers (Addition and Subtraction)

- Comparing and combining numbers through hundreds and thousands
- Comparing and combining decimals
- Using landmark numbers to find sums and differences
- Estimating and computing distances

Changes Over Time (Graphs)

- Examining situations and representations that show change
- Making graphs that show change over time
- Relating changes and the total
- Using curves to communicate changes
- Making, describing, and interpreting line graphs

Packages and Groups (Multiplication and Division)

- Factors and multiples
- Multiplying single-digit numbers
- Multiplying two-digit numbers
- Relating multiplication and division
- Representing division situations with notation

Sunken Ships and Grid Patterns (2-D Geometry)

- Locating points on the coordinate grid
- Naming points with ordered pairs
- Measuring distances on the coordinate grid
- Creating Logo procedures for rectangles and patterns

Three out of Four Like Spaghetti (Data and Fractions)

- Collecting and organizing categorical data
- Representing and analyzing data
- Describing data using fractions
- Using fractions to compare data

GRADE 4 UNIT-BY-UNIT CORRELATION TO NCTM STANDARDS

Mathematical Thinking at Grade 4 (Introduction)

Students develop and share strategies as they investigate numbers in the hundreds and ways to efficiently mentally operate on numbers. They investigate patterns in number and computation. In geometry, they work with patterns of symmetry, investigating both mirror and rotational symmetry.

Mathematical Emphases	Estimation	Number Sense and Numeration	Whole Number Operations	Whole Number Computation	Geometry and Spatial Sense	Measurement	Statistics and Probability	Fractions and Decimals	Patterns and Relationships
	5	6	7	8	9	10	11	12	13
Grouping things for more efficient counting	1:1	1:1 2:1			1:1				
Reordering numbers for more efficient mental arithmetic	TMM 1:4	1:1, 4 2:1 TMM	1:1, 4 2:1 TMM	1:4 TMM					
Finding how many more are needed	1:4 2:3-4	1:4 2:1,3-4	1:4 2:1,3-4	1:4 2:1,3-4					
Estimating how many hundreds in the total of a group of three-digit numbers	1:1 TMM	1:1 TMM	1:1 TMM	1:1					
Recognizing values of U.S. coins and grouping coins for more efficient counting		2:1,3-4 3:4-5	2:1,3-4 3:4-5	2:1,3-4 3:4-5					
Recognizing the decimal point on the calculator		2:1	2:1	2:1					
Using known answers to find other answers		3:3,4-5							3:3,4-5
Subtracting on a 300 chart and with a calculator		3:1-2	3:1-2	3:1-2					
Adding and subtracting multiples of ten	1:4	1:4 3:1-2, 4-5	1:4 3:1-2, 4-5	3:1-2, 4-5					
Distinguishing between geometric patterns and random designs					4:1, 2, 3-4				4:1, 2, 3-4
Distinguishing between mirror symmetry and rotational symmetry					4:1, 2, 3-4,5-6				4:1, 2, 3-4,5-6
Collecting, representing, and interpreting data							TMM		TMM
Estimating totals and differences	TMM	TMM	TMM					TMM	

TMM = Ten-Minute Math
1:2-3 = Investigation 1, Sessions 2-3

Arrays and Shares (Multiplication and Division)

Through activities, students develop a sense of what multiplication and division are and how these processes are related. They gain fluency with multiplication and division pairs and solve problems using their own strategies as well as by breaking problems into manageable components.

	NCTM Standards								
Mathematical Emphases	Estimation	Number Sense and Numeration	Whole Number Operations	Whole Number Computation	Geometry and Spatial Sense	Measurement	Statistics and Probability	Fractions and Decimals	Patterns and Relationships
	5	6	7	8	9	10	11	12	13
Using skip counting as a model for multiplication		1:1-2 3:2-4 TMM	1:1-2 2:1,2-3 TMM		2:2-3				1:1-2 3:2-4 TMM
Seeing multiplication as an accumulation of groups of a number		1:3 2:4-5 TMM	1:3 2:2-3, 4, 5-6 TMM	1:3 2:5-6	2:5-6				1:3 TMM
Looking for the multiplication patterns of numbers (including patterns of multiples highlighted on the 100 chart)		1:1-2,3	1:1-2,3 2:5-6 3:2-4 TMM	1:3 2:5-6 3:2-4 TMM					1:1-2, 3 2:5-6 3:2-4 TMM
Using known multiplication relationships to solve harder relationships		1:3 2:5-6 3:1,2-4	1:3 2:5-6 3:1	1:3 2:5-6 3:1	3:1				1:3 2:5-6 3:1
Using an array as a model for multiplication		2:2-3, 4, 5-6	2:1, 2-3, 4, 5-6	2:2-3, 4, 5-6	2:1, 2-3, 4, 5-6				2:2-3, 4, 5-6
Recognizing prime numbers as those that each have only one pair of factors and only one array		2:2-3	2:2-3		2:2-3				2:2-3
Understanding how division notation represents a variety of division situations (including sharing and partitioning situations)			2:7-8 3:2-4	2:7-8 3:2-4	2:7-8				2:7-8
Determining what to do with leftovers in division, depending on the situation			2:7-8 3:2-4	2:7-8 3:2-4				2:7-8	
Partitioning numbers to multiply them more easily (e.g., 7 x 23 can be 7 x 10 plus 7 x 10 plus 7 x 3)		3:1, 2-4, 5	3:1, 2-4, 5	3:1, 2-4, 5	3:1				3:1, 2-4, 5
Learning about patterns that are useful for multiplying by multiples of 10		3:2-4 TMM	3:2-4	3:2-4					3:2-4 TMM

TMM = Ten-Minute Math
1:2–3 = Investigation 1, Sessions 2–3

Seeing Solids and Silhouettes (3-D Geometry)

Students develop spatial visualization skills. They pictorially represent solid shapes, then build cube configurations from pictures, mental images, and different types of instructions.

	NCTM Standards								
	Estimation	Number Sense and Numeration	Whole Number Operations	Whole Number Computation	Geometry and Spatial Sense	Measurement	Statistics and Probability	Fractions and Decimals	Patterns and Relationships
Mathematical Emphases	5	6	7	8	9	10	11	12	13
Developing concepts and language needed to reflect on and communicate about spatial relationships in 3-D environments					1:1, 2 2:1-2				
Understanding standard drawings of 3-D cube configurations					1:1, 2 2:3 3:2-3	1:1			
Exploring spatial relationships between components of 3-D figures					1:1 2:3, 4				
Developing visualization skills		TMM			1:1, 2 2:3 3:1,2-3 TMM				
Starting to think about problems related to volume		1:1	1:1		1:1	1:1			
Understanding how 3-D geometric solids project shadows with 2-D shapes (e.g., how a cone can project a triangular shadow)					2:1-2,3				
Understanding geometric perspective					2:1-2, 3, 4				
Learning to visualize objects from different perspectives, then integrating views to form a mental model of the whole object					2:1-2, 3, 4 3:2-3				
Interpreting different types of instructions for building with cubes and evaluating the effectiveness of different forms of "how-to" instructions					3:1,2-3 4:1				
Integrating information given in separate views or presented verbally to form one coherent mental model of a cube figure					2:3 3:2-3				
Communicating effectively about three-dimensional objects					1:1 4:1				

TMM = Ten-Minute Math
1:2–3 = Investigation 1, Sessions 2–3

Landmarks in the Thousands (The Number System)

Students explore the structure of our number system. They work with factors and multiples of 100 and 1000, identify patterns, and use landmark numbers to solve addition and subtraction problems.

Mathematical Emphases	Estimation	Number Sense and Numeration	Whole Number Operations	Whole Number Computation	Geometry and Spatial Sense	Measurement	Statistics and Probability	Fractions and Decimals	Patterns and Relationships
	5	6	7	8	9	10	11	12	13
Finding and counting by factors of 100		1:1	1:1						1:1
Recognizing factor pairs (e.g., 4 rows of 25 cubes make 100; 25 rows of 4 cubes make 100)		1:2	1:2		1:2	1:2			1:2
Using landmarks to find differences between numbers under 100 (e.g., the difference between 48 and 100 is 52 because from 48 to 50 is 2, and then it's 50 more)		1:3	1:3						
Making conjectures about factors of 100		1:1							1:1
Using knowledge of the factors of 100 to explore multiples of 100 (e.g., if there are four 25's in 100, then there are eight in 200)		2:1, 2-4, 5	2:1, 2-4, 5						2:1, 2-4
Relating knowledge of factors to division situations and to standard division notations (e.g., 700 ÷ 20 means "How many 20's are in 700?" and can be solved by skip counting or reasoning)		2:1, 5	2:1, 5	2:5					2:1
Adding and subtracting multiples of 10 to numbers in the hundreds, and later, multiples of 100 to numbers in the thousands		2:2-4 3:2,3-5 4:1-3	2:2-4 3:2,3-5 4:1-3	2:2-4 3:3-5 4:1-3					2:2-4 3:3-5 4:1-3
Solving addition and subtraction problems by reasoning from known relationships		2:2-4 3:3-5	2:2-4 3:3-5	2:2-4 3:3-5					2:2-4 3:3-5
Reading, writing, and locating in sequence, numbers to 1000, and later, numbers in the thousands		3:1, 2 4:1-3							3:1
Getting a sense of the magnitude of multiples of 100 up to 1000	3:3-5	4:1-3							
Identifying and using important landmarks up to 1000, including the factors of 1000 and the multiples of those factors (e.g., 25, 50, 75, 100, 125, 150, 175, 200, . . .)	3:3-5	1:3 3:2,3-5	3:2						3:2
Developing strategies for adding and subtracting numbers in the hundreds		3:3-5	3:3-5	3:3-5					3:3-5
Estimating quantities up to 1000	3:3-5	3:3-5							
Getting a sense of the magnitude of 10,000 and understanding its structure (e.g., it can be constructed as 10 thousands or 100 hundreds)		4:1-3							4:1-3
TMM—Investigating the likelihood of events							2:1,5 TMM		
TMM—Counting to become familiar with multiples, factors, and multiplication patterns		3:3-5 4:1-3 TMM							

TMM = Ten-Minute Math
1:2–3 = Investigation 1, Sessions 2–3

Different Shapes, Equal Pieces (Fractions and Area)

Students explore fractions by dividing square areas into halves, fourths, and eighths and rectangular areas into thirds, sixths, and twelfths. They compare and order fractions, including fractions greater than one, and identify equivalent fractions.

Mathematical Emphases	Estimation	Number Sense and Numeration	Whole Number Operations	Whole Number Computation	Geometry and Spatial Sense	Measurement	Statistics and Probability	Fractions and Decimals	Patterns and Relationships
	5	6	7	8	9	10	11	12	13
Understanding that equal fractions of a whole have the same area but are not necessarily congruent					1:1,2-4 2:1-2	1:1,2-4		1:1,2-4 2:1-2	
Experiencing that cutting and pasting shapes conserves their area					1:1,2-4	1:1,2-4		1:1,2-4	
Becoming familiar with relationships among halves, fourths, and eighths, and then among thirds, sixths, and twelfths		1:2-4 2:1-2	1:5		1:2-4 2:1-2, 3, 4	1:2-4 2:3, 4		1:2-4,5 2:1-2, 3, 4	1:2-4,5
Knowing that equal fractions of different-sized wholes will be different in area					2:1-2	2:1-2		2:1-2	
Using different combinations to make a whole		2:3	2:3	2:3	1:5 2:3, 4			1:5 2:3, 4	1:5
Working with fractions that have numerators larger than one		2:3 3:1-2	2:3	2:3	2:3, 4 3:1-2			1:5 2:3, 4 3:1-2	1:5 2:4
Comparing any fraction to the landmarks 0, ½, 1, and 2		3:3			3:3	3:3		3:3	
Using both numerical reasoning and areas to order fractions (e.g., 4/9 is smaller than ½ because 2 x 4/9 = 8/9 which is less than 1)		3:3,4-5			3:3,4-5	3:3,4-5		3:3,4-5	3:3,4-5
Using the size of the numerator to compare fractions that have the same denominator and using the size of the denominator to compare fractions with the same numerator		3:4-5						3:4-5	
Understanding the fractions "missing one piece" are ordered inversely to the size of the missing piece (e.g., ⅔ is smaller than ¾ because the ⅓ missing is larger than the ¼ missing)					3:4-5			3:4-5	3:4-5
Identifying equivalent fractions					3:1-2	3:1-2		3:1-2	3:1-2
TMM—Using logical reasoning and relationships among numbers to guess a number		1:2-4 2:1-2 3:1-2 TMM	1:2-4 2:1-2 TMM	1:2-4 2:1-2 TMM				3:1-2 TMM	

TMM = Ten-Minute Math
1:2–3 = Investigation 1, Sessions 2–3

The Shape of the Data (Statistics)

Students record, represent, and analyze simple data sets. They organize data in working draft and then presentation form, and describe the shape of the data distribution.

					NCTM Standards				
	Estimation	Number Sense and Numeration	Whole Number Operations	Whole Number Computation	Geometry and Spatial Sense	Measurement	Statistics and Probability	Fractions and Decimals	Patterns and Relationships
Mathematical Emphases	5	6	7	8	9	10	11	12	13
Making quick sketches of the data to use as working tools during the analysis process		1:1,2-3				2:1,2-3	1:1,2-3 2:1 3:1		2:2-3 3:1
Describing the shape of the data, moving from noticing individual features of the data to describing the overall shape of the distribution	1:1	1:1, 2-3, 6-7				2:1	1:1,2-3 2:1, 4, 6-7 3:1	3:1	1:2-3 2:1
Defining the way data will be collected		1:2-3				2:1, 2-3	1:2-3 2:1,2-3 3:1,3-5		2:1
Summarizing to express what is typical of the data		1:2-3 2:1					1:2-3 2:1, 4, 6-7 3:1,3-5		1:2-3 2:4 3:1
Inventing ways, including representations, to compare two sets of data by describing the shape of the data and what's typical of the data		2:1				2:4	1:2-3 2:1, 2-3, 4		2:1, 2-3
Revising and refining sketches to make a presentation graph or chart								2:2-3 3:3-5	
Visualizing and estimating lengths and heights; using linear measure						2:1, 4	2:1, 2-3, 4		2:1
Using the median to describe a set of data to compare one data set to another		2:6-7	2:6-7	2:6-7		2:4,6-7	2:4,6-7		
Understanding that the median is only one landmark in the data							2:5,6-7		
Finding the median in a set of data arranged in numerical order (e.g., when students line up in order by height)		2:5,6-7				2:5,6-7	2:5,6-7		2:5
Finding the median in a set of data grouped by frequency (e.g., on a line plot or other graph)							2:6-7		
Carrying out all the stages of a data analysis investigation							3:1		
Choosing and refining a research question							3:1,3-5		
Viewing the data in several different ways; using quick sketches and other representations to organize and display the data							3:3-5		
TMM—Developing strategies for mental computations and judging the reasonableness of results	1:2-3 2:1 TMM	1:2-3 2:1 TMM						1:2-3 2:1 TMM	
TMM—Finding alternate paths to an answer		2:4 3:1-2 TMM	2:4 3:1-2 TMM	2:4 3:1-2 TMM					2:4 3:1-2 TMM

TMM = Ten-Minute Math
1:2–3 = Investigation 1, Sessions 2–3

Money, Miles, and Large Numbers (Addition and Subtraction)

Students add and subtract decimal numbers and numbers in the hundreds and thousands within the contexts of money and distance.

Mathematical Emphases	Estimation (5)	Number Sense and Numeration (6)	Whole Number Operations (7)	Whole Number Computation (8)	Geometry and Spatial Sense (9)	Measurement (10)	Statistics and Probability (11)	Fractions and Decimals (12)	Patterns and Relationships (13)
Estimating sums, including total amounts of money	1:1-2, 3, 4-5, 7-8 2:1-2 3:1,2-4	1:1-2, 3, 4-5, 7-8 3:1,2-4	1:1-2, 4-5,7-8 3:1,2-4	1:7-8 3:1,2-4					
Exploring strategies for comparing and combining numbers, through hundreds and thousands	1:1-2, 3, 4-5, 7-8 3:1,2-4	1:1-2, 3, 4-5, 6,7-8 3:1,2-4	1:1-2, 3, 4-5, 6,7-8 2:1-2 3:1,2-4	1:1-2, 3, 6, 7-8 3:1,2-4					
Using landmark numbers (multiples of 10 or 0.10 and 100 or 1.00) to compare and find differences between two quantities	1:1-2, 3,4-5 3:1,2-4	1:1-2, 3, 4-5 3:1,2-4	1:1-2, 3, 4-5 3:1,2-4	1:1-2 3:1,2-4					
Using standard addition and subtraction notation to record combining and comparing situations				1:6,7-8 3:2-4					
Using the calculator to solve problems and interpreting decimals on the calculator as amounts of money		1:4-5, 7-8	1:4-5, 7-8	1:4-5, 7-8				1:4-5	
Estimating local distances in miles and tenths of miles: developing a sense of about how long a mile and $\frac{1}{10}$ of a mile are	2:1-2,3	2:3			2:1-2,3	2:1-2,3			2:3
Comparing and combining decimal numbers and, later, quantities with decimal portions	2:1-2	2:1-2	1:6,7-8 2:1-2,4	1:6,7-8 2:1-2,4		2:4			
Seeing the relationships of decimal parts to the whole						2:1-2,3		2:1-2,3	
Measuring distances on maps using a scale	2:4 3:2-4	3:2-4	3:2-4	3:2-4		2:4 3:2-4		2:4	2:4 3:2-4
Becoming familiar with common decimal and fraction equivalents		2:1-2,3						2:1-2, 3, 4	
TMM—Considering whether events are likely or unlikely to occur							1:3 2:7-8 3:1 TMM		

TMM = Ten-Minute Math
1:2–3 = Investigation 1, Sessions 2–3

Changes Over Time (Graphs)

Students investigate change over time and ways to describe and represent changes. They explore continuous and discrete changes, and get an overall sense of change from a graph.

Mathematical Emphases	Estimation	Number Sense and Numeration	Whole Number Operations	Whole Number Computation	Geometry and Spatial Sense	Measurement	Statistics and Probability	Fractions and Decimals	Patterns and Relationships
	5	6	7	8	9	10	11	12	13
Deciding how to group data							1:1-2, 3-4		
Inventing representations of data, including graphs to show change over time						3:2	1:1-2, 3-4 3:1, 2, 6-7		3:1, 2, 6-7
Interpreting different kinds of graphs							1:1-2		
Developing a scale that includes all the data							1:3-4 3:3		
Establishing conventions for consistency							1:3-4		
Understanding how changes and total are related		1:5-6	1:5-6	1:5-6			1:5-6		
Developing strategies for solving missing-information problems when the information is missing from the beginning, middle, or end			1:5-6	1:5-6			1:5-6		
Writing missing-information problems			1:5-6	1:5-6					
Examining real situations and events that show change		3:2					2:1-2 3:1, 2	3:2	
Making and interpreting representations that show change		3:6-7				3:1, 2	2:1-2 3:1, 2, 6-7		3:2, 6-7
Distinguishing between representations of something that can change and representations that show changes							2:1-2		
Using curves to communicate information		3:2, 4, 5, 6-7				2:1-2 3:2, 4, 6-7	2:1-2 3:4, 5, 6-7		2:1-2 3:4, 6-7
Developing an understanding of the difference between continuous and discrete changes		3:2					3:2	3:2	3:2
Making, interpreting, and comparing line graphs		3:3, 4				3:2, 4	3:1, 2, 3, 4		
Integrating quantitative, qualitative, and graphical descriptions of the same data						3:4, 5	3:1, 4, 5		3:1, 4, 5
Making and interpreting different graphical shapes		3:4, 6-7				3:4, 5	1:3-4 3:1, 2, 4, 5, 6-7		3:1, 2, 4, 5, 6-7
TMM—Developing visual images of spatial representations		1:3-4, 2:1-2 TMM			1:3-4, 2:1-2 TMM				
TMM—Finding alternate ways to arrive at a solution		3:3, 4 TMM	3:3, 4 TMM	3:3, 4 TMM					

TMM = Ten-Minute Math
1:2–3 = Investigation 1, Sessions 2–3

Packages and Groups (Multiplication and Division)

Students continue their work with multiplication. They find multiples and factors, continue to learn single-digit multiplication pairs, and solve multi-digit multiplication and division problems.

Mathematical Emphases	Estimation (5)	Number Sense and Numeration (6)	Whole Number Operations (7)	Whole Number Computation (8)	Geometry and Spatial Sense (9)	Measurement (10)	Statistics and Probability (11)	Fractions and Decimals (12)	Patterns and Relationships (13)
Looking for and using multiplication patterns of numbers (e.g., identifying multiples of 5 by seeing that the units digit is either a 5 or a 0)		1:1-2 3:4-6	1:1-2 3:4-6	3:4-6	1:1-2				1:1-2 3:4-6
Finding multiples and becoming familiar with the multiples of larger numbers (e.g., skip counting by 2-digit numbers like 25)	1:4-5	1:3,4-5 3:4-6	1:3,4-5 3:4-6	1:4-5 3:4-6	1:4-5				1:3,4-5 3:4-6
Identifying factors of larger numbers (including triple-digit numbers)		3:7-8,9	1:4-5 3:7-8,9	1:4-5 3:7-8,9					3:7-8,9
Using familiar landmark numbers to solve problems (e.g., determining whether the solution is greater than 100, 200, 300, etc., or estimating 32 x 9 as 30 x 10 or 300)	2:2-3 3:4-6	2:1,2-3 3:4-6	2:1,2-3 3:4-6, 10	2:1 3:4-6, 10					2:1
Partitioning large numbers to multiply them more easily (e.g., 24 x 8 is thought of as 20 x 8 + 4 x 8)	2:2-3	2:1,2-3 3:3	2:1,2-3 3:3,4-6	2:1,2-3 3:3,4-6					2:1
Solving double-digit multiplication problems (e.g., 32 x 21)	2:2-3	2:2-3	2:2-3 3:4-6	2:2-3 3:4-6					2:2-3
Understanding how division notation can represent a variety of division situations, including sharing and grouping situations	3:3	3:1-2,3	3:1-2, 3, 4-6, 10	3:10					
Creating a context that is representative of a division equation (e.g., representing 152 ÷ 4 = 38 with 152 apples divided into 38 packages of 4)	3:3		3:1-2, 3, 10	3:1-2, 3, 10					
Using multiplication and division relationships in order to solve problems	3:4-6	3:3,4-6	3:1-2, 3, 4-6	3:1-2, 3, 4-6					
TMM—Describing features of data; interpreting and posing questions about data							1:4-5 TMM		
TMM—Recognizing and describing characteristics of numbers and relationships among numbers		3:3,7-8 TMM							

TMM = Ten-Minute Math
1:2–3 = Investigation 1, Sessions 2–3

Sunken Ships and Grid Patterns (2-D Geometry)

Students name and locate points on a coordinate grid with ordered pairs of numbers, both positive and negative, and measure distances on the grid. Students discuss properties of rectangles and write rectangle procedures for the computer using Geo-Logo™.

		NCTM Standards							
Mathematical Emphases	Estimation	Number Sense and Numeration	Whole Number Operations	Whole Number Computation	Geometry and Spatial Sense	Measurement	Statistics and Probability	Fractions and Decimals	Patterns and Relationships
	5	6	7	8	9	10	11	12	13
Using positive and negative coordinates to name and locate points on grids		1:2, 3-4			1:1, 2, 3-4	1:2, 3-4			1:3-4
Calculating distances on a grid based on paths along grid lines					1:1, 2, 3-4, 5-6	1:1, 2, 3-4, 5-6			1:5-6
Exploring numerical patterns that represent geometric situations		1:3-4			1:3-4, 5-6	1:3-4, 5-6			1:3-4, 5-6
Connecting visual and numerical descriptions of distances on a grid		1:5-6	1:5-6		1:3-4, 5-6	1:3-4, 5-6			
Applying knowledge of coordinates to locate points on a computer screen		1:3-4 2:2-3			1:3-4, 5-6 2:2-3	1:3-4, 5-6 2:2-3			
Describing geometric figures such as rectangles and squares in several ways					2:1, 6-7	2:1			
Understanding how Logo commands and patterns of commands reflect the properties of geometric figures					2:1, 4, 5, 6-7	2:1, 4, 5, 6-7			2:4
Creating and applying patterns and mental arithmetic strategies to solve turtle geometry problems		2:5	2:5		2:4, 5	2:4, 5			2:4
Using mirror and rotational symmetry to place rectangles on a grid and to design complex patterns of rectangles		2:2-3			2:2-3, 6-7, 8-9	2:2-3, 6-7, 8-9			2:2-3, 8-9
TMM—Relating the perimeter of a polygon to the length of its sides and using arithmetic operations in the context of perimeter		1:5-6 2:4 TMM	1:5-6 2:4 TMM		1:5-6 2:4 TMM	1:5-6 2:4 TMM			

TMM = Ten-Minute Math
1:2–3 = Investigation 1, Sessions 2–3

Three out of Four Like Spaghetti (Data and Fractions)

Students collect, describe, display, and compare categorical data. They classify the data in different ways and use fractions to describe and compare the categorizations.

Mathematical Emphases	Estimation	Number Sense and Numeration	Whole Number Operations	Whole Number Computation	Geometry and Spatial Sense	Measurement	Statistics and Probability	Fractions and Decimals	Patterns and Relationships
	5	6	7	8	9	10	11	12	13
Partitioning a group according to a rule		1:1					1:1	1:1	1:1
Finding familiar fractions ($\frac{1}{2}, \frac{1}{4}, \frac{1}{3}$) of a group	1:2	1:2				1:2	1:2	1:2	
Collecting, recording, and analyzing categorical (nonnumerical) data and describing data in terms of fractions	1:3 2:2	1:3					1:3 2:1, 2, 3, 5-7	1:3 2:2, 5-7	1:3 2:5-7
Using fractions to compare data from two groups, including two groups of different sizes	1:3	1:3					1:3	1:3 2:5-7	
Recognizing that fractions are always fractions of a particular whole								1:3	1:3
Estimating complex fractions with familiar fractions (e.g., $\frac{12}{25}$ is about $\frac{1}{2}$)	1:2 2:5-7	2:5-7					1:2 2:5-7	1:2 2:5-7	
Organizing data into categories and defining categories to accommodate additional data							2:1, 2, 3, 4, 5-7		
Making judgments about sets of categories							1:4 2:1, 2, 3,4 ,5-7	2:2, 3	2:3, 5-7
Representing categorical data, including use of bar graphs, and describing the data	2:2						1:4 2:1, 2, 5-7	2:2	
TMM—Making predictions about outcomes and exploring samples							1:3 2:2 TMM		

TMM = Ten-Minute Math
1:2–3 = Investigation 1, Sessions 2–3

GRADE 5 OVERVIEW OF CORRELATION TO NCTM STANDARDS

	NCTM Standards								
	Number and Number Relationships	Number Systems and Number Theory	Computation and Estimation	Patterns and Functions	Algebra	Statistics	Probability	Geometry	Measurement
Grade 5 Units	5	6	7	8	9	10	11	12	13
Mathematical Thinking at Grade 5 (Introduction, Landmarks in the Number System)	✔	✔	✔	✔				✔	
Picturing Polygons (2-D Geometry)	✔	✔	✔	✔					✔
Name That Portion (Fractions, Percents, and Decimals)	✔	✔	✔	✔		✔			
Between Never and Always (Probability)	✔	✔	✔			✔	✔		
Building on Numbers You Know (Computation and Estimation Strategies)	✔	✔	✔		✔				
Measurement Benchmarks (Estimating and Measuring)			✔			✔			✔
Patterns of Change (Tables and Graphs)	✔	✔	✔	✔	✔			✔	✔
Containers and Cubes (3-D Geometry: Volume)	✔		✔					✔	✔
Data: Kids, Cats, and Ads (Statistics)	✔	✔	✔			✔		✔	✔

Grade 5 Unit Topics

Mathematical Thinking at Grade 5 (Introduction, Landmarks in the Number System)

- Relationships among landmarks of 100, 1000, and 10,000
- Computational strategies that rely on landmarks up to 10,000
- Using number characteristics (multiple, factor, even, odd, prime, and square) to solve problems
- Exploring number composition through repeated addition, skip counting, finding factors and factor pairs, and using a calculator to check divisibility
- Magnitude of numbers through 10,000

Picturing Polygons (2-D Geometry)

- Properties of regular and nonregular polygons, particularly triangles and quadrilaterals
- Generating geometric figures with certain properties using Geo-Logo™, shape pieces, and coordinate grid paper
- Coordinate grids
- Relationships among turns, angles, and other characteristics of polygons
- Similar polygons

Name That Portion (Fractions, Percents, and Decimals)

- Equivalent fractions, decimals, and percents
- Relationships among fractions, from halves to twelfths
- Comparing and ordering fractions, decimals, and percents using a variety of models
- Choosing and using models and notation to compute with fractions, decimals, and percents
- Breaking fractions, decimals, and percents into familiar parts
- Planning, conducting, and presenting surveys

Between Never and Always (Probability)

- Verbal and numeric descriptions of probability
- Interpreting probability as a measure or quantity
- Using probability to select events most likely to occur
- Features of distributions of chance events including center, shape, and variability
- Fair (equal probability) games

Building on Numbers You Know (Computation and Estimation Strategies)

- Strategies for estimating and solving computation problems
- Modeling situations with the four basic operations
- Remainders
- Relationships between multiplication and division

Measurement Benchmarks (Estimating and Measuring)

- Measuring and estimating length, weight, volume, and time
- Sources of measurement error
- Developing and using benchmarks for length, weight, volume, and time to estimate, order, and compare
- Using and comparing metric and U.S. standard measures

Patterns of Change (Tables and Graphs)

- Building, extending, and describing tile patterns that grow according to regular number patterns
- Making, interpreting, and comparing tables, graphs, and stories that show accumulated distance and speed
- Relationships among distance, time, and speed
- Comparing graph shapes to describe rates of growth
- Relating position/time and velocity/time graphs

Containers and Cubes (3-D Geometry: Volume)

- Volume and units of volume
- Determining the volume of rectangular prisms, rooms, containers, and geometric solids
- Appropriate units of volume
- Relating the dimensions of a shape to its volume

Data: Kids, Cats, and Ads (Statistics)

- Collecting, compiling, organizing, representing, and analyzing data
- Relationships between variables in a data set
- Medians and other fractional parts of data sets
- Strategies for finding representative samples
- Making theories, statements, conclusions, and recommendations based on organized data
- Estimating the size of and finding equivalents to fractions in order to combine and compare them

GRADE 5 UNIT-BY-UNIT CORRELATION TO NCTM STANDARDS

Mathematical Thinking at Grade 5 (Introduction, Landmarks in the Number System)

Students are introduced to the methods and materials of the *Investigations* curriculum. They build an understanding of the factors and multiples of 10,000. They develop solutions to computations from their number sense and knowledge of the base ten system.

Mathematical Emphases	Number and Number Relationships	Number Systems and Number Theory	Computation and Estimation	Patterns and Functions	Algebra	Statistics	Probability	Geometry	Measurement
	5	6	7	8	9	10	11	12	13
Developing, discussing, and comparing strategies for solving problems about number relationships and computation	1:2,5-7 3:2-4,5 4:1	1:1-2 3:2-4,5	1:1-2 3:2-4,5 4:1						
Reasoning about and describing number characteristics and relationships such as multiple, factor, even, odd, prime, and square	1:5-7 4:2,3,4	1:3-4, 5-7 4:2,3,4						1:3-4	
Representing factor pairs as dimensions of a rectangular array		1:3-4 2:2-4,5							
Exploring materials that will be used as problem-solving tools, including calculators	1:1-2 4:2,3,4	1:1-7 4:2,3,4	1:1-2 4:2,3,4						
Communicating mathematical thinking through written and spoken language	1:1-2, 5-7 3:5 4:5-6	1:3-4,5-7 2:2-4 3:5 4:5-6	1:1-2 3:5 4:5-6						
Solving problems with one solution, more than one solution, and no solutions	1:5-7 4:2,3,4 4:5-6	1:5-7 4:2,3,4 4:5-6	4:2,3,4 4:5-6						
Using knowledge of landmarks up to 100 (including factors of 100 and multiples of them) to explore landmarks up to 1000, and using landmarks up to 1000 to explore landmarks up to 10,000	2:1,2-4 3:1,5 4:5-6	2:1,2-4 3:1,5 4:5-6	3:1 3:5	2:1,2-4 3:1,5					
Developing a variety of strategies for exploring number composition (e.g. repeated addition, skip counting, finding factors and factor pairs, using a calculator to check divisibility)	2:1 3:5 4:5-6	2:1,2-4 3:5 4:5-6	2:1,2-4 3:5 4:5-6						
Reading, writing, and ordering numbers to 1000 and 10,000	2:5	2:5		2:5					
Developing a sense of the magnitude of 1000 and 10,000	2:2-4,5 4:5-6	2:5 4:5-6							
Becoming familiar with skip-counting patterns leading to 1000 (e.g. sequences of multiples of 25, 50, and 75) and 10,000 (multiples of 250, 500, 750)	2:1,2-4 3:1 4:5-6		2:1 3:1	2:1 3:1 4:5-6					
Becoming familiar with factors and factor pairs of 1000 and 10,000	2:2-4 3:1 4:5-6	2:2-4 3:1 4:5-6	2:2-4 3:1 4:5-6	4:5-6				2:2-4,5	
Using knowledge of landmarks up to 10,000 (including factors of 1000 and multiples of those factors) to solve puzzles and problems	4:2,3,4, 5,6 3:1	3:1 4:2,3,4, 5-6	3:1 4:2,3,4, 5-6						
Developing mental multiplication and division strategies that rely on landmarks up to 10,000	3:2-4	3:2-4	3:2-4						

TMM = Ten-Minute Math
1:2–3 = Investigation 1, Sessions 2–3

Grade 5 Correlation to NCTM Standards 5–13

Mathematical Thinking at Grade 5 (Introduction, Landmarks in the Number System), cont.

	NCTM Standards								
	Number and Number Relationships	Number Systems and Number Theory	Computation and Estimation	Patterns and Functions	Algebra	Statistics	Probability	Geometry	Measurement
Mathematical Emphases	5	6	7	8	9	10	11	12	13
Developing mental and written strategies for finding sums and differences of 3- and 4-digit numbers	4:1,2,3,4	4:1,2,3,4	4:1,2,3,4						
TMM—Interpreting, posing questions about data, and using fractions to describe data						1:5-7 2:1 TMM			
TMM—Developing and analyzing concepts and language to reflect on and communicate about spatial relationships, shapes, patterns, and visual images	3:1 TMM							3:1 4:1 TMM	

TMM = Ten-Minute Math
1:2–3 = Investigation 1, Sessions 2–3

Picturing Polygons (2-D Geometry)

Students describe and create polygons on paper, with plastic shapes, and with Geo-Logo™. They investigate properties of triangles and quadrilaterals, and work with regularity and similarity.

NCTM Standards

Mathematical Emphases	Number and Number Relationships 5	Number Systems and Number Theory 6	Computation and Estimation 7	Patterns and Functions 8	Algebra 9	Statistics 10	Probability 11	Geometry 12	Measurement 13
Distinguishing between polygons and nonpolygons and between regular and nonregular polygons				1:1 3:1-2				1:1-2 3:1-3	3:1-3
Recognizing and naming polygons by number of sides								1:2-4 2:1	
Locating points on a coordinate grid				1:3-4 2:4-5				1:3-4 2:4-5	
Generating geometric figures with certain properties (including in a geometry computer environment)								1:3-4 2:4-7 3:1-2, 4-7	2:4-7 3:1-2
Sorting and classifying triangles and quadrilaterals, and developing vocabulary to describe special cases				2:1-3				1:2 2:1-5	2:1-5
Developing an understanding of parallel lines								2:1-7	
Distinguishing and seeing relationships between turns and angles			2:9 3:3					2:6-7,9 3:1-3	2:6-7,9 3:1-2
Using known angles to find the measures of others			2:8					2:1-3,8	2:8
Estimating and measuring the sizes of angles and turns			3:3					2:1-3, 6-9 3:1-3	2:1-3, 6-9 3:1-3
Finding the sizes and the sums of turns and angles in regular and nonregular polygons, and exploring the relationship to the total number of sides			3:1-3	3:1-3	3:1-3			3:1-3	3:1-3
Writing computer procedures to draw regular polygons and figures that are similar to a given figure	3:5-7							3:1-2, 4-7	3:1-2
Creating geometric patterns that grow in regular ways	3:4-7	3:4	3:4-7	3:4-7				3:4	3:4
Exploring connections between geometric and numerical patterns				3:1-7				3:1-7	
Exploring relationships among angles, line lengths, and areas of similar polygons	3:4-7	3:4	3:4-7	3:4-7				3:4-7	3:4-7
TMM—Relating factors to their multiples and developing number sense about multiplication and division relationships	1:2 2:4-5 TMM	1:2 2:4-5 TMM	1:2 2:4-5 TMM						

TMM = Ten-Minute Math
1:2-3 = Investigation 1, Sessions 2–3

Name That Portion
(Fractions, Percents, and Decimals)

Students use grids, arrays, number lines, clocks, and gender-participation surveys to make fraction, decimal, and percent comparisons and to solve computation problems. Through games and other activities, they develop familiarity with common fraction relationships.

Mathematical Emphases	Number and Number Relationships	Number Systems and Number Theory	Computation and Estimation	Patterns and Functions	Algebra	Statistics	Probability	Geometry	Measurement
	5	6	7	8	9	10	11	12	13
Interpreting everyday situations that involve fractions, decimals, and percents	1:1,2,7 2:9 3:1,7 4:1-4	1:1,2,7 2:9 3:1,7 4:1-4	1:7 2:9 3:7 4:1-4			3:1,7			
Using fractions and percents to name portions of groups	1:1-4,7 2:9 3:7 4:1-2, 5-7	1:1-4,7 2:9 3:7 4:1-2, 5-7	1:7 2:9 3:7 4:1-2, 5-7						
Breaking fractions, decimals, and percents into familiar parts	1:1-4,7 2:4-9 3:5-8	1:1-4,7 2:4-9 3:5-8	2:4-8						
Approximating data as familiar fractions and percents, and in circle graphs	1:1,2 4:1-7	1:1,2 4:1-7	1:1,2 4:1-7			4:3-7			
Identifying and using equivalent fractions, percents, and decimals	1:1-7 2:1-9 3:1,3-8 4:1-7	1:1-7 2:1-9 3:1,3-8 4:1-7	1:2-7 2:1-9 3:1,3-8 4:1-7						
Representing, comparing, and ordering fractions (common; mixed number; with numerators larger than 1; with different denominators), decimals, and percents using landmark numbers and visual models	1:2-7 2:1-9 3:1-8 4:1-7	1:2-7 2:1-9 3:1-8 4:1-7	1:2-7 2:1-9 3:1-8 4:1-7						
Choosing models and notations to compute with fractions, percents, and decimals	1:7 2:1-9 3:1-8 4:1-7	1:7 2:1-9 3:1-8 4:1-7	1:7 2:1-9 3:1-8 4:1-7						
Identifying and labeling fractions between 0 and 1 on a number line to make an array of fractions	2:4-5	2:4-5	2:4-5						
Finding patterns in an array of fraction number lines and in a decimal table	2:4-5 3:5-6			2:4-5 3:5-6					
Solving word problems and expressing answers to fit the context	1:1,7 2:9 3:7	1:1,7 2:9 3:7	1:1,7 2:9 3:7						
Finding decimals that are smaller than, larger than, or in between other decimals	3:3-6	3:3-6	3:3-6						
Planning and conducting surveys, and compiling, organizing, and communicating the results						4:1-7			
TMM—Finding ways to describe number relationships, including fraction notation, factor pairs, and equations	1:3-4 2:3,6 TMM	1:3-4 2:3,6 TMM			1:3-4 2:3,6 TMM				
TMM—Interpreting, posing questions about, and using fractions to describe data	3:2,5-6 TMM	3:2,5-6 TMM	3:2,5-6 TMM			3:2,5-6 TMM			

TMM = Ten-Minute Mat
1:2–3 = Investigation 1, Sessions 2–

Between Never and Always (Probability)

Students develop their probabilistic intuition by conducting experiments, analyzing the fairness of games, and comparing expectations to what actually happens.

Mathematical Emphases	Number and Number Relationships 5	Number Systems and Number Theory 6	Computation and Estimation 7	Patterns and Functions 8	Algebra 9	Statistics 10	Probability 11	Geometry 12	Measurement 13
Distinguishing certain events from those that are not, and distinguishing events with different probabilities							1:1-2		
Learning to associate the word *probability* with how likely something is to occur							1:1-2		
Matching verbal and numeric descriptions of probability	1:1-2	1:1-2					1:1-2		
Linking equivalent fractions, decimals, and percents	1:1-2	1:1-2					1:1-2		
Interpreting a probability as a measure or quantity	1:1-2	1:1-2, 3-4,5					1:1-2, 3-4,5		
Understanding that repeating a probability experiment can produce a variety of results						1:3-4	1:3-4,5 2:1-2		
Using probability to predict how often an event will happen and to select events most likely to occur		1:3-4, 5,7	1:5				1:3-4,5 2:1-2		
Plotting results of probability experiments on line plots and interpreting the data represented						1:3-4, 5,6	1:3-4, 5,6		
Comparing expected outcomes with actual outcomes						1:3-4, 5,6	1:3-4, 5,6		
Estimating probabilities from results of actual trials						1:6	1:6		
Inferring theoretical probabilities (e.g. $\frac{1}{4}$) from looking at spinners divided into sectors						1:3-4	1:3-4,5		
Predicting and analyzing features of distributions, including center and variability						1:3-4, 5,6	1:3-4, 5,6		
Identifying numbers in terms of multiples/factors, odds/evens, primes		1:7							
Adding probabilities of simple events (e.g. of rolling a die and getting 2) to obtain probabilities of types of events (rolling a die and getting an even number)		1:7	1:7				1:7		
Interpreting the fairness of a game as equal probability to win						2:1-2, 3,4-5	2:1-2, 3,4-5		
Breaking composite events into elementary events							1:7		
Developing systematic ways to generate a list of all possibilities							1:7 2:1-2		
Applying knowledge of probability to design a fair game, and writing directions others can follow							1:7 2:1-2, 4-5		
Distinguishing games of chance from games of skill							2:1-2, 3		

MM = Ten-Minute Math
2–3 = Investigation 1, Sessions 2–3

Between Never and Always (Probability), cont.

	NCTM Standards								
	Number and Number Relationships	Number Systems and Number Theory	Computation and Estimation	Patterns and Functions	Algebra	Statistics	Probability	Geometry	Measurement
Mathematical Emphases	5	6	7	8	9	10	11	12	13
Analyzing group data in terms of general features (e.g. center, spread)						2:3	2:3		
Appreciating that, even in fair games, variability in results can make a game appear unfair						2:3	2:1-2,3		
TMM—Approximating numbers, calculating mentally, and comparing numbers to find the closest answer	1:3-4 2:1-2 TMM		1:3-4 2:1-2 TMM						

TMM = Ten-Minute Math
1:2–3 = Investigation 1, Sessions 2–3

Building on Numbers You Know (Computation and Estimation Strategies)

Students invent and explain strategies for adding, subtracting, multiplying, and dividing that are based on their understanding of the numbers in the problems.

Mathematical Emphases	Number and Number Relationships 5	Number Systems and Number Theory 6	Computation and Estimation 7	Patterns and Functions 8	Algebra 9	Statistics 10	Probability 11	Geometry 12	Measurement 13
Skip counting by 2-, 3-, and 4-digit numbers (including landmark numbers)	1:1-5 4:1-2 5:4-6	1:1-5 4:1-2 5:4-6	1:1-5 4:1-2 5:4-6						
Relating skip counting to multiplication and division		1:1-5 4:1-2 5:4-6							
Finding and using patterns in sequences of multiples	1:1-5	1:1-5	1:1-5						
Reading, writing, and ordering large numbers, and approximating them to the nearest multiple of 100 or 1000	1:2,5-8 4:1-2 5:1-2, 4-6	1:2,5-8 4:1-2 5:1-2, 4-6							
Developing strategies for determining and comparing distances between numbers	1:2,6-8 5:1-2		1:2,6-8 5:1-2						
Using random digits to approximate 4- or 5-digit numbers	1:6-8	1:6-8							
Developing, recording, explaining, and comparing strategies for estimating and solving subtraction, multiplication, and division problems in more than one way	1:2-8 2:1-7 3:1-10 4:1-2 5:1-8	1:2-5 2:1-7 3:1-10 4:1-2 5:1-8	1:2-8 2:1-7 3:1-10 4:1-2 5:1-8						
Making sense of remainders in a variety of contexts	2:1-7 3:4-10 5:1-8	2:1-7 3:4-10 5:1-8	2:1-7 3:4-10 5:1-8						
Interpreting, recording, and using division and multiplication notation in a variety of situations			2:1-7 3:1-10 5:1-8		2:1-7 3:1-10 5:1-8				
Understanding and explaining the relationships among the four basic operations, and using those relationships to solve problems and model situations	1:1-8 2:1-3, 5-6 3:4-10 4:1-2 5:1-8	1:1-5, 6-8 2:1-3, 5-6 3:4-10 4:1-2 5:1-8	1:1-8 2:1-3, 5-6 3:4-10 4:1-2 5:1-8		1:1-8 2:1-3, 5-6 3:4-10 4:1-2 5:1-8				
Developing real-life meaning for quantities in the thousands, ten thousands, and hundred thousands, and beginning to acquire a sense of the size of 1,000,000	2:7 4:1-2 5:4-6								2:7
Breaking difficult computation problems into manageable parts	3:1-10 5:1-8	3:1-10 5:1-8	3:1-10 5:1-8						
Using a rectangular array model to represent factor pairs of numbers 10,000 and larger	4:1-2 5:4-6	4:1-2 5:4-6	4:1-2 5:4-6						
TMM—Visualizing ratios, making predictions, and exploring the relationship of a sample to its group	1:5 2:1-2					1:5 2:1-2	1:5 2:1-2		
TMM—Developing concepts and language to communicate about shapes, patterns, and visual images								3:1-3 5:1-2	

TMM = Ten-Minute Math
2–3 = Investigation 1, Sessions 2–3

Measurement Benchmarks (Estimating and Measuring)

Students estimate, take, and compare measurements using tools such as meter sticks, balance scales, liter measures, and timers. They compare metric and U.S. standard units of measure qualitatively, without arithmetic conversion.

Mathematical Emphases	NCTM Standards								
	Number and Number Relationships	Number Systems and Number Theory	Computation and Estimation	Patterns and Functions	Algebra	Statistics	Probability	Geometry	Measurement
	5	6	7	8	9	10	11	12	13
Using tools for measuring length, weight, volume, and time									1:1,3,4, 5-6,7 2:3,4 3:1
Recognizing metric and U.S. standard measurement units									1:1 2:1-2,4
Recognizing uses of benchmarks in estimation			2:3						1:2,3
Deciding when precise measurement is required and when estimates are sufficient			1:7-8			1:7-8			1:2-3, 7-8
Recognizing and explaining possible sources of measurement error			1:7-8			1:7-8			1:4,7-8
Comparing lengths expressed in different ways (e.g. meters and centimeters, meters and decimal fractions of a meter, meters and fractions of a meter)									1:4,5-6
Keeping track of and calculating total measurements			1:5-6, 7-8						1:3,4, 5-6,7-8
Developing benchmarks for and measuring distances of 100 meters			1:5-6						1:5-6
Comparing distances expressed in hundreds or thousands of miles or kilometers			1:7-8			1:7-8			1:7-8
Calculating approximate distances on maps			1:5-6, 7-8			1:7-8			1:5-6, 7-8
Exploring conversion relationships between metric and U.S. standard measures of weight and liquid						2:1-2			2:1-8
Ordering items by measures of weight and of liquid amount			2:1-2						2:1-2, 7-8
Gaining a sense of metric and U.S. standard measures of weight and capacity, and developing benchmarks for these measures									2:1,2,3, 4
Measuring with a liter measure marked in milliliters									2:4
Developing a sense of volume as the space something occupies or the capacity of a container									2:2,4,5
Reasoning about factors that influence capacity (e.g. features of a container's shape, such as height and width)									2:4
Beginning to develop the concept of density									2:5

TMM = Ten-Minute Math
1:2–3 = Investigation 1, Sessions 2–3

Measurement Benchmarks (Estimating and Measuring), cont.

Mathematical Emphases	Number and Number Relationships 5	Number Systems and Number Theory 6	Computation and Estimation 7	Patterns and Functions 8	Algebra 9	Statistics 10	Probability 11	Geometry 12	Measurement 13
Distinguishing between quantity and weight									2:5
Writing about weight, liquid capacity, and density									2:3,4,5
Using graphs to organize data and to determine typical data	2:7-8	2:7-8				2:7-8			2:7-8
Developing benchmarks for large numbers of pounds			2:7-8						2:7-8
Determining relative quantities: how many times as heavy or how many times as long as another object an object is			2:7-8						2:7-8
Developing vocabulary for units of time									3:1
Developing benchmarks for minutes and years									3:1,3
Timing in minutes and seconds									3:1
Collecting and analyzing data	3:1,2					3:1,2			3:1
Keeping track of computation in a multistep problem			3:2,3			3:2			3:2,3
TMM—Looking at the problem as a whole, looking at the largest part of each number first, and reordering or combining numbers to simplify it	1:2 1:5-6 TMM	1:2 1:5-6 TMM	1:2 1:5-6 TMM						
TMM—Using evidence and formulating questions to logically and systematically order, sort, and eliminate measurement units as possible solutions									2:4 3:1 TMM

TMM = Ten-Minute Math
1:2-3 = Investigation 1, Sessions 2-3

Patterns of Change (Tables and Graphs)

Students use number patterns, graphs, and other visual representations to analyze movement, races, and the growth of geometrical patterns of their own design. They relate these representations to each other and to changing action that they describe.

Mathematical Emphases	Number and Number Relationships	Number Systems and Number Theory	Computation and Estimation	Patterns and Functions	Algebra	Statistics	Probability	Geometry	Measurement
	5	6	7	8	9	10	11	12	13
Building tile designs that grow according to regular number patterns, continuing them, and predicting later steps of number patterns, tiles designs, and graphs	1:1-4	1:1-4	1:1-4	1:1-4	1:1-4			1:1-4	1:1-4
Comparing sequences of numbers and shapes of graphs	1:1-4	1:1-4	1:1-4	1:1-4	1:1-4	1:1-4			1:1-4
Comparing graph shapes to describe rates of growth				1:1-4 2:1-5 3:1-7	1:1-4 2:1-5 3:1-7				1:1-4 2:1-5 3:1-7
Exploring, reflecting on, and representing relationships among distance, time, and speed	2:2-5 3:1-6		2:2-5 3:1-6	2:1-5 3:1-7	2:1-5 3:1-7	2:1-5 3:1-7			2:1-5 3:1-7
Making, interpreting, and comparing tables, graphs, and stories that show accumulated distance and speed	2:2-5 3:2-6		2:2-5 3:2-6	2:1-5 3:2-6	2:2-5 3:2-6	2:2-5 3:2-6			2:2-5 3:2-6
Collecting and recording data in regular time intervals to analyze patterns of change						2:2			2:2
Exploring relationships between discrete and continuous descriptions of motion				2:2 3:2	2:2 3:2				
Comparing relative motion				3:1-6	3:1-6				
Relating number patterns to graphical shapes				1:1-4 2:1-5 3:1-7	1:1-4 2:1-5 3:1-7				
Illustrating relative change or motion in an animation				3:7					
TMM—Approximating numbers, calculating mentally, and comparing numbers to find the closest answer	1:2-4 2:1 TMM		1:2-4 2:1 TMM						
TMM—Attending to important features of a graph (e.g. relative height, slope) to imagine the stories behind graphs of change over time and to draw graphs to fit particular stories				3:1 TMM	3:1 TMM	3:1 TMM		3:1 TMM	

TMM = Ten-Minute Math
1:2–3 = Investigation 1, Sessions 2–

Containers and Cubes
(3-D Geometry: Volume)

Students explore the concept of volume. They develop strategies for finding the volumes of boxes, and they investigate volume relationships between pyramids and prisms and cylinders and cones.

Mathematical Emphases	Number and Number Relationships (5)	Number Systems and Number Theory (6)	Computation and Estimation (7)	Patterns and Functions (8)	Algebra (9)	Statistics (10)	Probability (11)	Geometry (12)	Measurement (13)
Seeing 3-D rectangular arrays of cubes in terms of congruent layers								1:1-2	1:1-2
Determining how many cubes fit in a rectangular box	1:1-4		1:1-4					1:1-4	1:1-4
Applying multiplication to find the number of cubes in a box	1:1-4		1:1-4					1:1-4	1:1-4
Determining the relationship between the number of cubes that fill a rectangular box and its dimensions	1:1-4		1:1-4					1:1-4	1:1-4
Organizing packages to fill rectangular boxes	2:1-5		2:1-5					2:1-5	2:1-5
Developing strategies for enumerating rectangular packages that fill boxes	2:1-5		2:1-5					2:1-5	2:1-5
Designing boxes to hold packages of different sizes	2:3-4		2:3-4					2:3-4	2:3-4
Understanding the relationship between the dimensions of a box and how many packages fill it	2:1-5		2:1-5					2:1-5	2:1-5
Understanding the concept of volume and units of volume								1:1-4 2:1-5 3:1-3 4:1-9	1:1-4 2:1-5 3:1-3 4:1-9
Seeing and using cubic centimeters as a unit for measuring volume (including nonrectangular solids)	3:1 4:4-9		3:1 4:4-9					3:1 4:4-9	3:1 4:4-9
Deciding on, constructing, and visualizing appropriate units of volume for measuring large-scale spaces								3:2-3	3:2-3
Understanding characteristics of units of volume, such as shape and size								3:1-3 4:1-9	3:1-3 4:1-9
Developing, using, describing, and justifying methods of determining volume	1:1-4 2:1-5 3:1-3 4:1-9		1:1-4 2:1-5 3:1-3 4:1-9					1:1-4 2:1-5 3:1-3 4:1-9	1:1-4 2:1-5 3:1-3 4:1-9
Comparing the volume of one room to another	3:3		3:3					3:3	3:3
Exploring volume relationships among different containers and among solids, particularly those with the same base and height	4:2-9		4:2-9					4:1-9	4:1-9
Using geometric solids to design models and to determine their volume	4:7-9		4:7-9					4:7-9	4:7-9
TMM—Becoming familiar with multiplication patterns, relationships between factors and their multiples, and relationships between multiplication and division	1:1-2 2:3-4 TMM	1:1-2 2:3-4 TMM							
TMM—Using evidence and formulating questions to logically and systematically order, sort, and eliminate measurement units as possible solutions									4:1 TMM

TMM = Ten-Minute Math
1:2–3 = Investigation 1, Sessions 2–3

Data: Kids, Cats, and Ads (Statistics)

Students examine and compare data sets, including data sets with several variables. They learn about selecting a reasonable and fair sample. They use fractions between 0 and 1 to describe probabilities.

Mathematical Emphases	Number and Number Relationships (5)	Number Systems and Number Theory (6)	Computation and Estimation (7)	Patterns and Functions (8)	Algebra (9)	Statistics (10)	Probability (11)	Geometry (12)	Measurement (13)
Finding medians and other fractional parts of data sets	1:1-4 2:1-3 3:1-4 5:3-5		1:1-4 2:1-3 3:1-4 5:3-5			1:1-4 2:1-3 3:1-4 5:3-5			
Making theories, statements, conclusions, and recommendations based on organized data	1:2-4 3:2-3 4:3		1:2-4 3:2-3 4:3			1:1-4 2:1-2 3:2-4 4:3 5:1,3-5			
Using data characteristics to identify data sets, to describe numerical and categorical variables, and to compare a sample to a larger population						1:4 2:1-3 3:2-4 4:3 5:1,3-5			
Collecting, organizing, and collating data, and making line plots and tables to examine and compare data sets						1:1-4 2:1-3 3:1-4 4:2-3 5:2-5			1:1 2:1
Framing questions about variables in a data set, and using representations and descriptions to answer them						2:2-3 5:1-5			
Using a computer and database tool to enter, analyze, and examine data in a computer database						2:3 5:3-5			
Comparing and adding fractions using numerical reasoning and visual representations and by converting unfamiliar fractions to more familiar fractions	1:1-4 2:1-3 3:1-4 4:1-3 5:3-5	3:1-4	1:1-4 2:1-3 3:1-4 4:1-3 5:3-5					4:1-3	4:1-3
Finding equivalents among fractions, decimals, and percents, and using them to compare data from a sample with a target fraction	3:1-4 4:3 5:3-5		3:1-4 4:3 5:3-5			3:2-4 4:3 5:3-5			4:3
Learning what a sample is, learning what some of the factors (including size) that make a sample reasonable are, and discovering why a large sample tends to reflect a population better than a small sample						2:1 3:1-4 4:1-3 5:2			
Developing strategies to find representative samples						3:2-4 4:2-3 5:2			
Formulating, testing, defining, and refining survey questions, and designing a survey						5:1-2			
TMM—Reasoning about place value, and developing strategies for comparing numbers and the distance between numbers	1:2-3 2:2 TMM	1:2-3 2:2 TMM	1:2-3 2:2 TMM						
TMM—Organizing and finding the number of cubes that fill (and squares that cover) simple 3-D solids			3:2-3 4:2 TMM					3:2-3 4:2 TMM	3:2-3 4:2 TMM